Rapid Assessment Program

22

RAP Bulletin of Biological Assessment

A Marine Rapid Assessment of the Raja Ampat Islands, Papua Province, Indonesia

Sheila A. McKenna, Gerald R. Allen, and Suer Suryadi, Editors

Center for Applied Biodiversity Science (CABS)

Conservation International (CI)

University of Cenderawasih

Indonesian Institute ofSciences (LIPI)

Australian Institute of Marine Science

Western Australian Museum

RAP Working Papers are published by:
Conservation International
Center for Applied Biodiversity Science
Department of Conservation Biology
1919 M Street NW, Suite 600
Washington, DC 20036
USA
202-912-1000 telephone
202-912-9773 fax
www.conservation.org
www.biodiversityscience.org

Editors: Sheila A. McKenna, Gerald R. Allen, and Suer Suryadi
Design/Production: Glenda P. Fábregas
Production Assistant: Fabian Painemilla
Maps: Conservation Mapping Program, GIS and Mapping
Laboratory, Center for Applied Biodiversity Science at Conservation International
Cover photograph: R. Steene
Translations: Suer Suryadi

ISBN 1-881173-60-7
Library of Congress Card Catalog Number 2001098383

RAP Bulletin of Biological Assessment was formerly RAP Working Papers. Numbers 1-13 of this series were published under previous title.

Suggested citation: McKenna, S.A., G.R. Allen, and S. Suryadi (eds.). 2002. A Marine Rapid Assessment of the Raja Ampat Islands, Papua Province, Indonesia. RAP Bulletin of Biological Assessment 22. Conservation International, Washington, DC.

Funding for this Marine RAP study and publication was generously provided by David and Lucile Packard Foundation, the Henry Foundation, and the Smart Family Foundation Inc.

♺ Printed on recycled paper.

Table of Contents

Participants

Gerald R. Allen, Ph. D. (Ichthyology and
Science Team Leader)
Conservation International
1919 M St., N.W., Suite 600
Washington, DC 20036 USA
Mailing address:
1 Dreyer Road
Roleystone, WA 6111
Australia
Fax: (618) 9397 6985
Email: tropical_reef@bigpond.com

Jabz Amarumollo (Community Liaison Team)
Conservation International
Jalan Bhayangkara I, No. 33
Jayapura, Indonesia 99112
Email: jzmarllo@n2nature.com

Paulus Boli MSc. (Cenderawasih University,
Manokwari, Coral Reef Ecologist)
Cenderawasih University
Manokwari, Papua Province, Indonesia
Email: paul_boli@mailexcite.com

Mohammed Farid (Community Liaison Team)
Conservation International
Jalan Bhayangkara I, No. 33
Jayapura, Indonesia 99112
Email: ci-irian@jayapura.wasantara.net.id

La Tanda BSc. (Reef Fisheries)
Development Center for Oceanology (LIPI)
Biak Field Station
Biak, Papua Province
Indonesia

Douglas Fenner, Ph. D. (Reef Corals)
Australian Institute of Marine Sciences
P.M.B. No. 3
Townsville, Queensland 4810,
Australia
Email: d.fenner@aims.gov.au

Sheila A. McKenna, Ph. D. (Reef Ecology,
RAP Survey Team Leader)
Conservation Biology Department
Conservation International
1919 M St., N.W., Suite 600
Washington, DC 20036 USA
Email: s.mckenna@conservation.org

Roger Steene (Photographer)
P.O. Box 188
Cairns, Queensland 4870
Australia

John E. N. Veron, Ph. D. (Reef Corals)
Australian Institute of Marine Sciences
P.M.B. No. 3
Townsville, Queensland 4810
Australia
Email: j.veron@aims.gov.au

Fred E. Wells, Ph. D. (Malacology)
Department of Aquatic Zoology
Western Australian Museum
Francis Street
Perth, WA 6000
Australia
Email: wellsf@museum.wa.gov.au

Organizational Profiles

Conservation International

Conservation International (CI) is an international, non-profit organization based in Washington, DC. CI acts on the belief that the Earth's natural heritage must be maintained if future generations are to thrive spiritually, culturally, and economically. Our mission is to conserve biological diversity and the ecological processes that support life on earth and to demonstrate that human societies are able to live harmoniously with nature.

Conservation International
1919 M St., N.W., Suite 600
Washington, DC 20036 USA
(202) 912-1000 (telephone)
(202) 912-1030 (fax)
http://www.conservation.org

Conservation International (Indonesia)
Jalan Taman Margasatwa 61
Jakarta, Indonesia 12540
(62-21) 7883-8624/26 (telephone),
(62-21) 7800265 (fax)
http:// www.ci-indonesia@conservation.org

Conservation International (Papua Province)
Jalan Bhayangkara I, No. 33
Jayapura, Papua Province 99112
Indonesia (62-967) 523-423 (telephone and fax)

University of Cenderawasih (UNCEN)

University of Cenderawasih serves as a center of excellence for Papuan students and provides a range of educational services to the community. UNCEN's mission is to train and enhance the technological and human resources of Papua to the benefit of the Papuan people and community. The focus has been on agriculture, economics, and fisheries. The Faculty of Agriculture is based at Manokwari and includes the department of Fishery and Marine Science.

University of Cenderawasih
Kampus Waena Jayapura
Jalan Sentani Abepura
Jayapura, Papua Province 99351
Indonesia
(62-967) 572-108 (telephone)
(62-967) 527-102 (fax)

Papua State University

The State University of Papua was formerly a branch of UNCEN, but was recently granted status as a separate institution. An integral part of its program involves agriculture, forestry, and fisheries.

Papua State University
Jalan Gunung Salju Amban
Manokwari, Papua Province
Indonesia
(62-986) 211-974 (telephone)
(62-986) 211-455 (fax)

Indonesian Institute of Sciences (LIPI)

The Indonesian Institute of Sciences is a non-departmental institution that reports directly to the President of Indonesia. The main tasks of LIPI are to assist the President in organizing research and development, and to provide guidance, services, and advice to the government on national science and technology policy. In order to accomplish its main tasks LIPI was assigned the following functions:

1. To carry out research and development of science and technology.
2. To encourage and develop science consciousness among the Indonesian people.
3. To develop and improve cooperation with national as well as international scientific bodies in accordance with the existing laws and regulations.
4. To provide the government with the formulation of national science policy.

Research and Development Center for Oceanology (RDCO)

RDCO is one of the main branches in the LIPI organization and is responsible for all aspects of marine science, including oceanography, marine biology, marine resources, and conservation technology for the general marine environment.

Research and Development Center for Oceanology
Jalan Pasir Putih I, Ancol Timur
P.O. BOX 4801/JKTF
Jakarta 11001, Indonesia
(62-21) 683-850 (telephone)
(62-21) 681-948 (fax)
http://www.oseanologi.lipi.go.id

Australian Institute of Marine Science

The mission of the Australian Institute of Marine Science (AIMS) is to generate the knowledge to support the sustainable use and protection of the marine environment through innovative, world-class scientific and technological research. It is an Australian Commonwealth Statutory Authority established by the Australian Institute of Marine Science Act of 1972 in recognition of a national need to manage Australia's marine environment and marine resources.

Australian Institute of Marine Science
Cape Ferguson, Queensland
PMB No 3, Townsville MC QLD 4810
(61-7) 4753-4444 (telephone)
(61-7) 4772-5852 (fax)
http://www.aims.gov.au

Western Australian Museum

The Western Australian Museum was established in 1891 and its initial collections were geological, ethnological and biological specimens. The 1960s and 1970s saw the addition of responsibility to develop and maintain the State's anthropological, archaeological, maritime archaeological and social and cultural history collections. The collections, currently numbering over two million specimens/artefacts, are the primary focus of research by the Museum's own staff and others. The aim is to advance knowledge on them and communicate it to the public through a variety of media, but particularly a program of exhibitions and publications.

Western Australian Museum
Francis Street
Perth, WA 6000
Australia
(61-8) 9427-2716 (telephone)
(61-8) 9328-8686
http://www.museum.wa.gov.au

Acknowledgments

This Marine RAP survey was financed by generous donations from the David and Lucile Packard Foundation, the Henry Foundation, and the Smart Family Foundation Inc.

We are very grateful to the Rector of the University of Cenderawasih, Mr. F. A. Wospakrik, and the Rector of the University of Papua, Dr. F. Wanggai for supporting this project, and providing the necessary permits and excellent counterparts. Similarly, we thank Dr. Kurnaen Sumadiharga, Director of the Research and Development Center for Oceanology of LIPI for his continued support of CI RAP surveys. The survey would not have been possible without the additional support of the Research Institute of the University of Cenderawasih, and the Biak Research Station of the Development Center for Oceanology.

We also appreciate the assistance of the Museum of Zoology (LIPI), particularly Dr. Siti N. Prijono (Director), Ristiyanti M. Marwoto (invertebrates), Ike Rachmatika (fishes), and Agus Tjakrawidjaja (fishes).

We also thank the following government staff for providing permits and sharing data: John Piet Wanane (Head, Regency of Sorong), Joseph Kbarek (Head, Regency Planning Agency), Constant Karel Sorondanya (Head, Nature Conservation Agency), Ahmad Fabanyo (Head, Dinas Fisheries), A. Rahman Adrias (Head, Dinas Tourism), S. Banjarnahor (Head, Dinas Trade), and Mr. Faisal (Head, Sorong Police Station). In addition, the Indonesian Department of Immigration kindly issued permits that enabled our RAP scientists to perform their survey and training duties.

We are grateful to the people of the Raja Ampat Islands who allowed us to conduct this survey and extended their wonderful hospitality. We thank the Kepala Desa and people of the following villages for their assistance and sharing their knowledge: Waiweser, Arefi, Yansawai, Marandan Weser, Sapokren, Yenbeser, Friwen, Yenbuba, Waiweser, Yenbekwan, Yenwaupnoor, Sawinggrai, Kapisawar, Arborek, Lopintol, Wawiyai, Kabui, Waifoi, Fam, Mutus, Miosmanggare, Manyaifun, Selpele, and Salio. We also express our gratitude to Taher Arfan (Head of Kepulauan Raja Ampat Adat Council), Fatah Abdullah (Head of Kecamatan Samate), Octavianus Mayor (Head of Kecamatan Waigeo Selatan), and our guide Pak Mayor from Yenbuba village, who accompanied us on our visits to various villages.

Yuli Supriyanto and Maisyie helped us to obtain permits and to gather information in Sorong. We were capably assisted by CI-Indonesia staff, including the Director Jatna Supriatna, Ermayanti, Myrna Kusumawardhani, Mira Dwi Arsanty, and Hendrite Ohee.

Thanks are due Max Ammer, owner of Irian Diving, and his staff, for providing crucial logistic assistance during the RAP, and for sharing their extensive knowledge of the underwater attractions of the region. Additionally, thanks to Max Ammer and the Raja Ampat Research and Conservation Center for help with final map editing. We also thank the staff of P.T. Cendana Indopearls, particularly Project Manager Joseph Taylor and Assistant Manager David Schonell, for providing accommodation during our stay at Alyui Bay on Waigeo Island. Mark Allen assisted with color scanning and prepared the layout for the color pages appearing in this report.

Ringkasan Eksekutif

Pendahuluan

Laporan ini memaparkan hasil penilaian lapangan secara cepat di Kepulauan Raja Ampat, Indonesia, yang terletak di paling ujung barat Propinsi Papua dulu bernama Irian Jaya. Kepulauan ini terdiri dari beberapa pulau besar dan bergunung-gunung, yaitu Waigeo, Batanta, Salawati dan Misool serta ratusan pulau-pulau kecil di sekitarnya. Daratan dan lautan di sekelilingnya mencakup luas sekitar 43.000 km^2. Total populasi penduduk adalah 48,707 atau 7 jiwa/ km^2 berdasarkan sensus terakhir tahun 1998. Pulau-pulau ini merupakan bagian dari "segitiga karang" (Coral Triangle) yang terdiri dari Indonesia, Filipina, Malaysia, Papua New Guinea, Jepang dan Australia. Kawasan tersebut mendukung kehidupan keanekaragaman hayati laut terkaya di dunia, yang umumnya berpusat di habitat-habitat karang yang luas, bakau dan padang lamun.

Survai ini dilakukan oleh Marine Rapid Assessment Program (RAP) Conservation International (CI) bekerjasama dengan Universitas Cenderawasih dan Pusat Penelitian dan Pengembangan Oseanologi-Lembaga Ilmu Pengetahuan Indonesia (P3O-LIPI).

Gambaran Umum Marine RAP

Tujuan Marine RAP adalah untuk menghasilkan dan menyebarluaskan informasi keanekaragaman hayati di daerah pesisir dan laut secara cepat untuk kepentingan konservasi, dengan sebagian fokus untuk merekomendasikan prioritas pembentukan kawasan konservasi dan pengelolaannya. Marine RAP mengirim tim multidisiplin ilmu yang terdiri dari ahli-ahli kelautan dan sumberdaya pesisir untuk menilai tingkat keanekaragaman hayati dan peluang konservasi di areal yang telah ditentukan. Melalui inventarisasi bawah laut selama tiga minggu, survei Marine RAP menghasilkan daftar-daftar spesies yang merupakan indikator kekayaan biologi secara keseluruhan, mencatat beberapa parameter untuk menilai kualitas/kesehatan ekosistem secara keseluruhan. Pada setiap survei, RAP juga memperhatikan dan menilai kebutuhan penduduk lokal, yang kemudian dimasukkan sebagai bagian dari rekomendasi akhir.

Dengan membandingkan hasil-hasil dari beberapa survei, Marine RAP akhirnya difokuskan untuk memastikan bahwa perwakilan dari sampel keanekaragaman hayati laut akan dikonservasi di kawasan lindung dan melalui berbagai upaya konservasi.

Kepulauan Raja Ampat

Laut di sekitar Kepulauan Raja Ampat sangat kaya dengan organisme laut dan dihuni oleh terumbu karang paling asli di Indonesia. Walaupun daerah yang disurvei jarang penduduknya (sekitar 7.700 jiwa), dapat ditemukan tanda-tanda kerusakan habitat yang tampaknya dilakukan oleh orang luar daerah yang sudah mempraktekkan pengeboman dan peracunan ikan. Penebangan liar juga terlihat dalam kawasan cagar alam di Pulau Waigeo. Jelas sekali sangat dibutuhkan upaya konservasi untuk melindungi ekosistem laut yang rentan dan menjamin adanya pemanfaatan sumberdaya yang berkelanjutan bagi generasi mendatang.

Survei Raja Ampat

Survei Marine Rap di Kepulauan Raja Ampat dilakukan di 45 lokasi selama 15 hari (27 Maret –10 April 2001). Secara umum lokasi yang akan dikunjungi telah ditentukan sebelumnya untuk memaksimalkan keragaman habitat, sehingga memungkinkan diperolehnya daftar spesies keragaman hayati yang maksimal. Pada tiap lokasi, inventarisasi bawah air dilakukan pada tiga kelompok satwa yang merupakan indikator keanekaragaman terumbu karang secara menyeluruh, yaitu karang scleractinian, moluska dan ikan karang. Pengamatan tambahan dilakukan untuk mengetahui kondisi lingkungan di tiap lokasi, termasuk penilaian terhadap berbagai parameter ancaman. Pengamatan dan data dari kegiatan perikanan karang juga dilakukan.

Daerah survei (lihat peta) mencakup sekitar 6.000 km², meliputi karang-karang di Selat Dampier antara Batanta Utara dan Waigeo. Areal survei juga mencakup Pulau Pam (penduduk setempat menyebutnya pulau Pam) dan kelompok pulau Batang Pele, ujung barat Waigeo, termasuk Teluk Alyui, Pulau Kawe dan Wayag yang jaraknya tidak jauh dari timur laut Waigeo. Ke 45 lokasi dicapai dengan perahu bermotor, berangkat dari base camp di Pulau Kri dan teluk Alyui. Lokasi dua terdiri dari dua habitat sehingga dipisahkan dengan tanda 2a, karang tepi dan 2b karang laguna tersembunyi di pulau Kri. Dengan demikian total jumlah lokasi survei adalah 45 walaupun jumlah yang disebutkan adalah 44.

Ringkasan Hasil

Terumbu karang di Kepulauan Raja Ampat memiliki keanekaragaman hayati yang luar biasa dan umumnya dalam kondisi fisik yang baik. Namun telah ada tanda-tanda kerusakan yang mengkhawatirkan, terutama akibat praktek penangkapan ikan yang merusak. Catatan hasil dari survei ini adalah :

- *Karang* : tercatat 456 spesies karang keras, yang berarti lebih dari setengah jumlah karang di dunia. Tak satupun tempat dengan luas area yang sama memiliki jumlah spesies sebanyak ini.

- *Moluska* : Keragamannya tergolong tinggi dengan 699 spesies. Jumlah ini melampaui semua hasil survei RAP sebelumnya di sekitar kawasan, termasuk Papua New Guinea dan Filipina.

- *Ikan karang*: ditemukan 828 spesies, meningkatkan jumlah total spesies ikan yang diketahui di kepulauan ini menjadi 972 spesies. Teknik extrapolasi dengan menggunakan enam famili indikator kunci menunjukkan bahwa di kawasan ini diharapkan terdapat sedikitnya 1.084 spesies.

- *Perikanan karang* : terdapat 196 spesies, mewakili 59 genus dan 19 famili yang dikategorikan sebagai spesies ikan target untuk konsumsi. Dugaan rata-rata total biomasa untuk lokasi-lokasi di Kepulauan Raja Ampat jauh lebih besar dibandingkan lokasi-lokasi

lain di kawasan "Coral Triangle" termasuk Propinsi Milne Bay (Papua New Guinea), Kepulauan Togean-Banggai (Indonesia) dan kepulauan Calamianes (Filipina).

- *Kondisi karang* : Berdasarkan Indeks Kondisi Karang CI, diketahui bahwa 60 % dari terumbu karang yang disurvei dalam kondisi baik atau sangat bagus. Lokasi-lokasi tersebut memiliki kombinasi keragaman karang dan ikan yang terbaik, yang relatif bebas dari gangguan dan penyakit. Sebaliknya, 17% terumbu karang tergolong dalam kondisi jelek, tetapi terbatas pada teluk tersembunyi yang tingkat pengendapan lumpurnya tinggi.

Rekomendasi Konservasi

Kepulauan Raja Ampat menyokong kehidupan biota laut yang kaya dan beragam. Terumbu karang dan ikan-ikan sangat kaya, bahkan mungkin jumlah spesiesnya terbanyak dibandingkan kawasan lain yang sama luasnya di dunia. Kepulauan ini juga memiliki pemandangan bawah air dan daratan yang sangat indah. Walaupun kebanyakan lokasi itu berada dalam kawasan cagar alam, tetap ditemukan tanda-tanda kerusakan habitat, khususnya akibat ulah penangkap ikan illegal yang menggunakan bom dan sianida. Selain itu, ikan kerapu dan Napoleon merupakan target perusahaan perikanan komersil sehingga jumlahnya menurun drastis. Berdasarkan temuan tersebut, tampak jelas diperlukan upaya-upaya konservasi yang akan melindungi kawasan dengan sumberdaya biologi yang unik ini. Berdasarkan survei RAP ini, kami sampaikan beberapa rekomendasi sebagai berikut:

1. **Melakukan kampanye penyadaran lingkungan.** Penduduk lokal perlu menyadari keunikan hidupan liar di sekitar mereka dan ketergantungannya pada habitat alami, juga manfaat konservasi dan konsekwensinya jika tidak ada tindakan yang dilakukan. Hal ini dapat dicapai dengan berbagai cara seperti memasukkan materi dalam kurikulum SMP dan SMU, mengundang pembicara dalam pertemuan di kota, poster, video dan publikasi-publikasi bergambar.

2. **Meningkatkan partisipasi masyarakat dalam perencanaan dan pengelolaan konservasi.** Penduduk lokal memiliki peluang yang sangat bagus untuk melakukan dan mengelola inisiatif konservasi yang akan berperan penting untuk memelihara keanekaragaman hayati laut di perairan sekitarnya. Masyarakat harus bekerjasama untuk mewujudkan tujuan bersama dari konservasi terumbu karang

dalam jangka panjang. Pembentukan dewan konservasi yang terdiri dari orang tua yang dipercaya dan dihormati dari seluruh desa dalam kawasan ini akan menjadi ajang komunikasi yang sangat baik.

3. **Mengadakan program-program di masyarakat untuk meningkatkan partisipasi dalam kegiatan konservasi.** Partisipasi masyarakat dapat didorong melalui bantuan dana untuk (melalui lembaga-lembaga pemerintah, perusahaan dan lsm) berbagai kegiatan di masyarakat seperti bantuan pendidikan, pelayanan kesehatan dan perbaikan gereja.

4. **Mengadakan program –program pengembangan alternatif ekonomi berkelanjutan untuk menggantikan panangkapan ikan illegal.** Jika penduduk tidak memperoleh pendapatan dari menangkap ikan, maka harus ada alternatif yang berkelanjutan untuk menghasilkan uang. Aktivitas yang mungkin dikembangkan adalah ekowisata dan aktifitas lain yang terkait. Pengembangan ekowisata terbatas merupakan cara yang sangat bagus untuk melaksanakan konservasi di tingkat lokal. Masyarakat dapat memperoleh keuntungan finansial melalui pekerjaan yang berhubungan dengan wisata sekaligus berperan aktif menjaga terumbu karang. Terumbu karang yang kondisinya dijaga baik akan terus mendatangkan turis dan penduduk lokal akan memperoleh manfaat jika terumbu terus mendukung kebutuhannya untuk sumber daya laut.

5. **Mengembangkan inisiatif konservasi darat dan laut secara bersamaan.** Kepulauan Raja Ampat memberikan peluang langka untuk mengembangkan program konservasi darat dan laut. Ekosistem darat dan laut berhubungan erat di kawasan ini dan dampak dari daratan berpengaruh langsung pada habitat laut

6. **Mengkaji ulang tata batas cagar alam yang sudah ada.** Batas-batas yang sudah ada perlu ditinjau kembali untuk memastikan batas-batasnya dan efektif untuk melindungi perwakilan habitat utama di darat dan laut. Setiap usaha harus dibuat untuk mengubah "perlindungan di atas kertas" menjadi suaka yang dikelola baik dan diawasi oleh polisi hutan setempat.

7. **Mengontrol atau mengurangi aktivitas ilegal yang berdampak negatif bagi ekosistem alam.** Pengrusakan sumber daya alam tak terkendali dan penangkapan ikan yang berlebihan merupakan masalah di seluruh Indonesia. Akibatnya, diperlukan kepastian hukum, khususnya di tingkat lokal yang mencakup semua aspek dari pengrusakan lingkungan dan perikanan. Praktek-praktek penangkapan ikan yang merusak seperti penggunaan sianida dan bom adalah ilegal. Namun demikian, upaya menghentikan praktek tersebut sebenarnya tidak terjadi di kawasan seperti Kep. Raja Ampat. Masalah ini merajalela di seluruh Indonesia dan dapat diatasi jika upaya konservasi yang benar-benar efektif dilaksanakan. Pemerintah pusat dan daerah perlu mengalokasikan dana untuk kapal patroli, personil terlatih, dan sumber daya lainnya. Selain itu, penegakan hukum yang efektif harus didukung oleh hukuman yang setimpal dalam bentuk denda yang tinggi, penyitaan kapal dan peralatan penangkapan ikan, dan atau ancaman penjara. Penebangan kayu ilegal di dalam kawasan konservasi juga merupakan masalah. Berbagai ancaman bagi lingkungan pesisir berasal dari darat. Penebangan yang tidak terkontrol tidak hanya menghilangkan sumber daya alam yang berharga, tapi erosi dari lokasi penebangan menghasilkan endapan lumpur yang berpengaruh langsung pada terumbu karang.

8. **Mendukung penelitian yang sangat penting bagi perencanaan konservasi lingkungan laut.** Mengingat keanekaragaman hayati laut dan darat yang sangat tinggi, maka diperlukan penelitian lanjutan yang mendalam, khususnya yang berhubungan dengan biota laut yang langka dan hampir punah. Pembangunan stasiun biologi dan dukungan dana bagi mahasiswa akan sangat membantu penelitian-penelitian yang diperlukan.

9. **Meningkatkan pengumpulan data-data penting untuk perencanaan konservasi laut.** Data utama biologi dan pendukung non-biologi sangat penting untuk merancang strategi konservasi yang efektif. Tampaknya diperlukan serangkaian lokakarya, dimana kelompok para ahli dan stakeholdernya menelaah informasi yang tersedia untuk menghasilkan kesepakatan berupa strategi yang dapat dilaksanakan. Hasil penting dari proses ini adalah teridentifikasinya kesenjangan informasi dan usulan untuk mengisi kekosongan informasi itu.

10. **Mengadakan program pemantauan lingkungan jangka panjang.** Masyarakat lokal perlu dilatih untuk memonitor terumbu karang secara berkala. Hal ini mungkin dapat dicapai melalui kerjasama dengan universitas di Papua dan LSM konservasi.

11. **Mengadakan pelatihan menyelam bagi staf universitas lokal dan organisasi konservasi.** Masih sedikit penyelam terlatih yang bekerja untuk LSM dan universitas di propinsi Papua. Akibatnya, terdapat keterbatasan minat untuk melakukan konservasi laut. Sangat diperlukan promosi nilai-nilai konservasi laut oleh ahli biologi dari Papua. Salah satu cara terbaik untuk memperbaiki kelemahan ini adalah melatih lebih banyak penduduk lokal untuk menyelam, yang akan membantu meningkatkan penghargaan terhadap lingkungan bawah laut.

12. **Mengadakan rapid assessment survei.** Survei 2001 merupakan upaya awal yang sangat baik, tetapi masih diperlukan survei tambahan. Khususnya, perlu dilakukan survei di Misool, Salawati dan Waigeo timur, kawasan yang tidak dikunjungi pada RAP sebelumnya. Selain itu juga sangat baik untuk melakukan satu kali atau lebih survei RAP terpadu yang menggabungkan komponen laut, darat, akuatik, dan sosial/ekonomi.

Executive Summary

Introduction

This report presents the results of a rapid field assessment of the Raja Ampat Islands, Indonesia, which lie immediately off the extreme western tip of Papua Province, also known as Irian Jaya. The group consists of several large, mountainous islands including Waigeo, Batanta, Salawati, and Misool, and hundreds of small satellite islands. The land and surrounding sea occupy approximately 43,000 km². Total population of the archipelago is 48,707 or 7 persons per km² of land, according to the last census in 1998. The islands form an integral part of the "coral triangle," composed of Indonesia, Philippines, Malaysia, Papua New Guinea, Japan, and Australia. This region supports the world's richest marine biodiversity, mostly concentrated in extensive coral reef, mangrove, and seagrass habitats.

The survey was implemented by the Marine Rapid Assessment Program (RAP) of Conservation International (CI) in collaboration with the University of Cenderawasih and the Research and Development Center for Oceanology, a branch of the Indonesian Institute of Sciences (LIPI).

Overview of Marine RAP

The goal of Marine RAP is to rapidly generate and disseminate information on coastal and near-shore shallow-water marine biodiversity for conservation purposes, with a particular focus on recommending priorities for conservation area establishment and management. Marine RAP deploys multi-disciplinary teams of marine scientists and coastal resource experts to determine the biodiversity significance and conservation opportunities of selected areas. Through underwater inventories generally lasting three weeks, Marine RAP surveys produce species lists that serve as indicators of overall biological richness, as well as recording several measurements to assess overall ecosystem health. During each survey, RAP supports parallel assessments of local human community needs and concerns, which become incorporated into the final recommendations.

By comparing the results obtained from many surveys, Marine RAP is ultimately focused on ensuring that a representative sample of marine biodiversity is conserved within protected areas and through other conservation measures.

Raja Ampat Islands

The seas surrounding the Raja Ampat Islands are exceedingly rich for marine organisms and harbor some of the most pristine reefs in Indonesia. Although the survey area is sparsely populated (about 7,700 residents), there are disturbing signs of habitat destruction, apparently due to encroachment by outsiders, who have introduced blast and cyanide fishing. Illegal logging was also observed within the gazetted nature reserve on Waigeo Island. There is clearly an urgent need for conservation initiatives in order to protect fragile marine ecosystems and to insure sustainable resources for future generations.

The Raja Ampats Survey

The Marine RAP survey of the Raja Ampat Islands assessed 45 sites over a 15-day period (27 March-10 April 2001). General site areas were selected prior to the actual survey in order to maximize the diversity of habitats visited, thus facilitating a species list that incorporates maximum biodiversity. At each site, an underwater inventory was made of three faunal groups selected to serve as indicators of overall coral reef biodiversity: scleractinian corals, molluscs, and reef fishes. Additional observations were made on the environmental condition of each site, including evaluation of various threat parameters. Observations and data on reef fisheries were also gathered.

The survey area (see map) covered approximately 6,000 km², encompassing reefs of the Dampier Strait between northern Batanta and Waigeo. The area also included the Fam (local people recognized it as Pam) and Batang Pele Island groups, the westernmost tip of Waigeo, including Alyui Bay, as well as Kawe and the Wayag Islands, lying a short distance to the northwest of Waigeo. The 45 survey sites were reached by motor boats, operating from base camps at Kri Island and Alyui Bay. Site two consisted of two habitats and was split. The sites are denoted as 2a for the fringing reef and 2b for sheltered lagoon reef at Kri Island. Therefore the total number of sites surveyed is 45 although the sites listed by number go to 44.

Summary of Results

Reefs of the Raja Ampat Islands harbor excellent biodiversity and are mainly in good physical condition. However, there are disturbing signs of degradation, primarily as a result of destructive fishing practices. Notable results from the survey include:

- *Corals*: 456 species of hard corals were recorded, which is more than half of the world's total. No other area of comparable size has this many species.

- *Molluscs*: Diversity was comparatively high with 699 species. This total surpasses those from past RAP surveys in surrounding regions, including Papua New Guinea and the Philippines.

- *Reef Fishes*: A total of 828 species were recorded, raising the total known from the islands to 970 species. An extrapolation technique utilizing six key indicator families reveals that at least 1,084 species can be expected to occur in the area.

- *Reef Fisheries:* A total of 196 species, representing 59 genera and 19 families, were classified as target species for reef fisheries. The mean total biomass estimate for sites in the Raja Ampat Islands is considerably greater than for other previously sampled areas in the "coral triangle" including Milne Bay Province (Papua New Guinea), Togean-Banggai Islands (Indonesia), and Calamianes Islands (Philippines).

- *Coral Condition:* Using CI's Reef Condition Index, it was noted that 60% of surveyed reefs were in good or excellent condition. These are sites with the best combination of coral and fish diversity and are relatively free of damage and disease. In contrast, 17% of reefs were considered to be in poor condition, but these were mainly confined to sheltered bays with high levels of silting.

Conservation Recommendations

The Raja Ampat Islands support a rich and varied marine fauna. Corals and fishes are particularly rich, with perhaps the greatest number of species than any other place in the world of similar size. The islands also possess extraordinary underwater and above-water scenery. Although much of the area lies within a gazetted nature reserve (cagar alam), there are disturbing signs of recent habitat destruction, particularly by illegal fishers who use explosives and cyanide. In addition, commercial fishing ventures are targeting large groupers and Napoleon Wrasse, and stocks appear to be dwindling rapidly. There is clearly a need for conservation initiatives that will protect the regions unique biological resources. As a result of our RAP survey we make the following specific recommendations:

1. **Implement an environmental awareness campaign.** Local residents need to become aware of the uniqueness of their special wildlife and its dependence on particular natural habitats, as well as the advantages of conservation and the consequences if no action is taken. This can be achieved in a variety of ways including primary and secondary school curricula, guest speakers at town meetings, posters, videos, and illustrated publications.

2. **Promote community participation in conservation planning and management.** Local communities have a wonderful opportunity to implement and manage conservation initiatives that will play a critical role in maintaining marine biodiversity in surrounding waters. Communities need to work together to achieve the common goal of long-term reef conservation. The formation of a conservation council of trusted and respected elders representing all villages in the area would greatly facilitate communication.

3. **Establish community outreach programs to provide extra incentives for participation in conservation activities.** Community participation could be encouraged by establishing and helping to finance (through government agencies, private corporations, and NGOs) various outreach programs that involve educational assistance, health care, and church improvements.

4. **Establish programs to develop sustainable economic alternatives to replace illegal fishing.** If villagers are denied an income from fishing there must be sustainable alternatives to earn cash. Possible activities include eco-tourism and related activities. Limited development of ecotourism is an excellent method to implement conservation at the local level. Communities can reap financial benefits through tourism-related employment and also play an active part in conserving reefs. Reefs that are maintained in good condition will continue to draw tourists, and local communities will naturally benefit if the reefs continue to sustain their needs for marine resources.

5. **Develop terrestrial and marine conservation initiatives concurrently.** The Raja Ampat Islands afford the rare opportunity to develop terrestrial and marine conservation programs. Land and sea ecosystems are intimately linked in this area and terrestrial impacts have direct consequences on marine habitats.

6. **Review boundaries of existing wildlife reserves.** Current boundaries need to be reviewed to insure they can be justified and are effective for protecting a representative cross-section of all major marine and terrestrial habitats. Every effort should be made to convert so called "paper parks" to meaningful reserves that are properly managed and patrolled by resident rangers.

7. **Control or eliminate illegal activities that negatively impact natural ecosystems.** Indiscriminate destruction of natural resources and over-fishing are problems throughout Indonesia. Consequently, it may be necessary to enact more precise laws, particularly at the local level, covering all aspects of fishing and environmental destruction. Destructive fishing practices such as the use of cyanide and dynamite are illegal. However enforcement of these activities is virtually non-existent in areas such as the Raja Ampat Islands. This problem is rampant throughout Indonesia and needs to be addressed if truly effective conservation practices can be implemented. Local and national governments need to allocate funds for patrol boats, trained personnel, and other resources. Additionally, effective enforcement needs to be backed up by adequate penalties in the form of heavy fines, confiscation of boats and fishing equipment, and/or jail sentences. Illegal logging within designated nature reserves also poses a problem. A variety of threats to coastal environments originate from land-based sources. Uncontrolled logging not only depletes

valuable natural resources, but the erosion of logged sites contributes to harmful silt deposition that directly affects coral reefs.

8. **Facilitate studies that are essential for planning the conservation of marine environments.** Given the extraordinary biodiversity of both marine and terrestrial systems there is a need for continued in-depth studies, particularly with regards to potentially rare and endangered marine wildlife. The establishment of a biological field station and financial support of university students would greatly facilitate the necessary studies.

9. **Promote collection of data essential for marine conservation planning.** A host of biological and supporting non-biological data are essential in designing an effective conservation strategy. It may prove worthwhile to convene a series of workshops in which a group of relevant experts and stakeholders review existing information to achieve consensus on a workable strategy. Important results of this process would be the identification of information gaps and proposals for how to fill these gaps.

10. **Establish a long-term environmental monitoring program.** Local communities should be trained to periodically monitor their reef resources. This could perhaps be achieved through collaboration with Papuan universities and conservation NGOs.

11. **Provide dive training for staff of local universities and conservation organizations.** There are relatively few trained divers working for NGOs and universities in Papua Province. Consequently, there is limited enthusiasm for marine conservation. There is a genuine need for promotion of marine conservation values by Papuan biologists. One of the best ways to remedy this shortcoming is to train more local people to dive, which will foster a greater appreciation for the undersea environment.

12. **Conduct additional rapid assessment surveys.** The 2001 survey forms an excellent starting point, but more surveys are required. In particular, there is a need for surveys at Misool, Salawati, and eastern Waigeo, areas that were not visited during the current RAP. There is also excellent scope for one or more integrated RAP surveys that incorporate marine, terrestrial, aquatic, and social/economic components.

Overview

Introduction

The Raja Ampat Islands, situated immediately west of the Birdshead Peninsula, are composed of four main islands (Misool, Salawati, Batanta, and Waigeo) and hundreds of smaller islands, cays, and shoals. Much of the area consists of gazetted wildlife reserve (cagar alam), but there remains a critical need for biological surveys. Delegates at the January 1997 Conservation Priority-setting Workshop on Biak unanimously agreed that the Raja Ampats are a high-priority area for future RAP surveys, both terrestrial and marine. The area was also identified as the number one survey priority in Southeast Asia at CI's Marine RAP Workshop in Townsville, Australia, in May 1998. Due to its location near the heart of the "Coral Triangle" (the world's richest area for coral reefs encompassing N. Australia, Indonesia, Philippines, and Papua New Guinea) coupled with an amazing diversity of marine habitats, the area is potentially the world's richest in terms of marine biodiversity.

The area supports some of the richest coral reefs in the entire Indonesian Archipelago. The sparsely populated islands contain abundant natural resources, but unfortunately are a tempting target for exploitation. The islands have long enjoyed a form of natural protection due to their remote location, but as fishing grounds have become unproductive in areas to the west, the number of visits by outside fishing vessels has increased. Particularly over the past two to three years, there has been a noticeable increase in the use of explosives and cyanide by both outsiders and local people.

This report presents the results of a Conservation International Marine RAP (Rapid Assessment Program) survey of marine biodiversity in the Raja Ampat Islands, focusing on selected faunal groups, specifically reef-building (scleractinian) corals, molluscs, and fishes.

Additional chapters present the results of fisheries and reef condition surveys, as well as a study of marine resource use by local communities. The purpose of this report is to document local marine biodiversity and to assess the condition of coral reefs and the current level of fisheries exploitation in order to guide regional planning, marine conservation, and the use of sustainable marine resources.

Marine RAP

There is an obvious need to identify areas of global importance for wildlife conservation. However, there is often a problem in obtaining the required data, considering that many of the more remote regions are inadequately surveyed. Scarcity of data, in the form of basic taxonomic inventories, is particularly true for tropical ecosystems. Hence, Conservation International has developed a technique for rapid biological assessment. The method essentially involves sending a team of taxonomic experts into the field for a brief period, often 2–4 weeks, in order to obtain an overview of the flora and fauna. Although most surveys to date have involved terrestrial systems, the method is equally applicable for marine and freshwater environments.

One of the main differences in evaluating the conservation potential of terrestrial and tropical marine localities involves the emphasis placed on endemism. Terrestrial conservation initiatives are frequently correlated with a high incidence of endemic species at a particular locality or region. Granted other aspects need to be addressed, but endemism is often considered as one of the most important criteria for assessing an area's conservation worth. Indeed, it has become a universal measure for evaluating and comparing conservation "hot spots." In contrast, coral reefs and other tropical marine ecosystems frequently exhibit relatively low levels of endemism. This is particularly true throughout the "coral triangle" (the area including northern Australia, the Malay-Indonesian Archipelago, Philippines, and western Melanesia), considered to be the world's richest area for marine biodiversity. The considerable homogeneity found in tropical inshore communities is in large part due to the pelagic larval stage typical of most organisms. For example reef fish larvae are commonly pelagic for periods ranging from 9 to 100 days (Leis, 1991). A general lack of physical isolating barriers and numerous island "stepping stones" have facilitated the wide dispersal of larvae throughout the Indo-Pacific.

The most important feature to assess in determining the conservation potential of a marine location devoid of significant endemism is overall species richness or biodiversity. Additional data relating to relative abundance are also important. Other factors requiring assessment are more subjective and depend largely on the observer. Obviously, extensive biological survey experience over a broad geographic range yields the best results. This enables the observer to recognize any unique assemblages within the community, unusually high numbers of normally rare taxa, or the presence of any unusual environmental features. Finally, any imminent threats such as explosive fishing, use of cyanide, over-fishing, and nearby logging activities need to be considered.

Reef corals, fishes, and molluscs are the primary biodiversity indicator groups used in Marine RAP surveys. Corals provide the major environmental framework for fishes and a host of other organisms. Without reef-building corals, there is limited biodiversity. This is dramatically demonstrated in areas consisting primarily of sand, rubble, or seaweeds. Fishes are an excellent survey group as they are the most obvious inhabitants of the reef and account for a large proportion of the reef's overall biomass. Furthermore, fishes depend on a huge variety of plants and invertebrates for their nutrition. Therefore, areas rich in fishes invariably have a wealth of plants and invertebrates. Molluscs represent the largest phylum in the marine environment, the group is relatively well known taxonomically, and they are ecologically and economically important. Mollusc diversity is exceedingly high in the tropical waters of the Indo-Pacific, particularly in coral reef environments. Gosliner et al. (1996) estimated that approximately 60% of all marine invertebrate species in this extensive region are molluscs. Molluscs are particularly useful as a biodiversity indicator for ecosystems adjacent to reefs where corals are generally absent or scarce (e.g. mud, sand, seagrass beds, and rubble bottoms).

It was decided at the Marine RAP Workshop in Townsville, Australia (May 1998) that CI would focus its survey activities on the "Coral Triangle," because this is the world's richest area for coral reef biodiversity and also its most threatened. Accordingly the Marine RAP program has completed surveys at Milne Bay, Papua New Guinea in 1997 (Werner and Allen, 1998) and 2000 (Allen et al., in press), Calamianes Islands, Philippines in 1998 (Werner and Allen, 2000), and the Togean and Banggai Islands, Indonesia in 1998 (Allen et al., in press).

Historical Notes

The Raja Ampat Islands were one of the first areas in the East Indies to attract the attention of European explorers and naturalists. The French frigate L'Uranie under the command of Captain Freycinet visited western New Guinea and the Raja Ampats in 1819–1820. The surgeon-naturalists Quoy and Gaimard, who accompanied the expedition, published records of approximately 30 fish species following their return to France in 1824. Another French ship, the corvette La Coquille commanded by Captain Duperrey, visited Waigeo in 1823, followed by a second visit by Quoy and Gaimard aboard L'Astrolabe in 1826. These visits resulted in the description of about 40 additional fishes. Quoy and Uranie islands, which lie off the northwestern coast of Waigeo, bear testimony to these early expeditions.

Following the early French visits there was scant scientific interest in the area until the explorations of the famous British naturalist Alfred Russell Wallace. *The Malay Archipelago* (Wallace, 1869) gives an excellent account of his visit to Waigeo. After an epic 18 day voyage utilizing a flimsy sailing canoe, he reached Waigeo on 4 July 1860. Wallace spent nearly three months there, but did not record any significant observations of marine life. Rather, he was tirelessly occupied with the task of collecting birds and insects. Nevertheless, his accounts of the landscape and people of Waigeo are fascinating reading. In some respects, there have been few changes since Wallace's visit 141 years ago. For example, after negotiating the narrow channel (RAP Site 5) leading into Kabui Bay, Wallace noted in his book: we emerged into what seemed a lake, but which was in fact a deep gulf having a narrow entrance on the south coast. This gulf was studded along its shores with numbers of rocky islets, mostly mushroom shaped, from the water having worn away the lower part of the soluble coralline limestone, leaving them overhanging from ten to twenty feet. Every islet was covered with strange-looking shrubs and trees, and was generally crowned by lofty and elegant palms, which also studded the ridges of the mountainous shores, forming one of the most singular and picturesque landscapes I have ever seen.

Surprisingly, there has been little interest in the marine biology of the Raja Ampat Islands over the past century. Although Dutch scientists have long been familiar with the area, most of their studies and collections focused on terrestrial and freshwater organisms. The small amount of marine research mainly involved fishes and is outlined elsewhere (see Chapter 3 in this report).

Physical Environment

The Raja Ampat Islands are situated immediately west of the Papua mainland, between 0°20' and 2°15' S latitude, and 129°35' and 131°20' E longitude. The Archipelago and surrounding seas occupy approximately 40,000 km². The diverse array of unspoiled coral reefs and superb above-water scenery combine to produce one of the world's premier tropical wildlife areas. Seas are exceptionally calm for most of the year due to the prevailing pattern of light winds and sheltering influence of large high islands. The area is a natural wonderland punctuated by an endless variety of islands from coconut-studded coral cays that scarcely rise above sea level to the spectacularly steep rain-forested slopes of Batanta and Waigeo, soaring to an elevation of 600–1000 m. Marine navigation charts reveal there are at least 1,500 small cays and islets surrounding the four main islands. Perhaps the most spectacular aspect of the above-water scenery is the "drowned karst topography" characterized by hundreds of limestone islets that form a seemingly endless maze of "forested beehives and mushrooms" (especially well developed on Waigeo Island at Kabui Bay and at the Wayag Islands).

The area experiences a typical monsoon regime of winds and rainfall. The dry season extends from October to March. The highest rainfall is generally between April and September, although June and July are generally the wettest. Average rainfall during the dry season is about 17 cm per month and about 27 cm per month during the wet season. Winds are generally from the southeast between May and October, and mainly from the northwest between December and March. During November, April, and May, which are transitional periods, winds are light and variable.

Maximum daily tide fluctuation is approximately 1.8 m, with an average daily fluctuation of about 0.9–1.3 m. Periodic strong currents are common throughout the area, especially in channels between islands. Sea temperatures during the survey period were generally 27–28°C and severe thermoclines or areas of upwelling were not encountered.

Marine environments in the Raja Ampat Islands are incredibly diverse and include extensive coral reefs, mangroves, and sea grass beds. Coral reefs are mainly of the fringing or platform variety. Fringing reefs are highly variable with regards to exposure, ranging from open sea situations to highly sheltered bays and inlets. The northern coast of Batanta and western end of Waigeo in particular are strongly indented with an abundance of sheltered bays. Mayalibit Bay, the large inlet separating East and West Waigeo, is essentially a marine lake with a narrow channel at its south eastern extremity. It is mainly bordered by mangroves, but there is limited reef development at the southern end and in the channel, where currents are frequently severe.

Socio-economic Environment

The Raja Ampat Archipelago is part of the Sorong Regency, which is composed of five districts: Salawati, Samate, Misool, South Waigeo, and North Waigeo. The population consists of 48,707 residents (17,516 families) inhabiting 89 villages, with approximately seven people per km². The inhabitants are mainly of Papuan origin, although there is a significant Indonesian community on Misool (not visited during the present survey).

The actual survey area includes 23 villages (see chapter 6) with a total population of about 5,726, (see chapter 6) ranging in size from Arborek (98 people) to Fam (785). More than 90% of the adult population of this area is engaged in sustenance-level fishing. At most villages there is relatively little commercial activity, although some people collect sea cucumbers (holothurians) that are sold to merchants in Sorong, the nearest large population center on the mainland. Fishers from at least seven villages are currently using cyanide-containing chemicals to catch Napoleon Wrasse and large groupers. The chemicals are provided by Sorong merchants, who pay very low prices for the illegally captured fishes. Many villagers expressed concern about various illegal fishing methods, but admitted it was one of the few ways they could earn extra cash.

Survey Sites and Methods

General sites were selected by a pre-survey analysis that relied on literature reviews, nautical charts (particularly British Admiralty charts 3248, 3744, and 3745), and consultation with Max Ammer, owner and operator of Irian Diving. In addition G. Allen was familiar with the area, having made two previous visits in connection with fresh water surveys. Detailed site selection was accomplished upon arrival at the general area, and was further influenced by weather and sea conditions.

At each site, the Biological Team conducted underwater assessments that produced species lists for key coral reef indicator groups. General habitat information was also recorded, as was the extent of live coral cover at several depths. The main survey method consisted of direct underwater observations by diving scientists, who recorded species of corals, molluscs, and fishes. Visual transects were the main method for recording fishes and corals in contrast to the method for recording molluscs, which relied primarily on collecting live animals and shells (most released or discarded after identification). Relatively few specimens were preserved for later study and these were invariably species that were either too difficult to identify in the field or were undescribed. Further collecting details are provided in the chapters dealing with corals, molluscs, and fishes.

Table 1. Summary of survey sites for Marine RAP survey of the Raja Ampat Islands.

No.	Date	Location	Coordinates
1	27/3/01	W. Mansuar Island	0° 36.815' S, 130° 33.538' E
2a	27/3/01	Cape Kri, Kri Island	0° 33.470' S, 130° 41.362' E
2b	31/3/01	Cape Kri Lagoon	0° 33.380' S, 130° 41.234' E
3	27/3/01	S Gam Island	0° 30.761' S, 130° 39.409' E
4	28/3/01	N Kabui Bay, W. Waigeo	0° 18.761' S, 130° 38.581' E
5	28/3/01	Gam-Waigeo Passage	0° 25.570' S, 130° 33.796' E
6	28/3/01	Pef Island	0° 27.030' S, 130° 26.444' E
7	29/3/01	Mios Kon Island	0° 29.901' S, 130° 43.531' E
8	29/3/01	Mayalibit Bay, Waigeo	0° 17.851' S, 130° 53.595' E
9	29/3/01	Mayalibit Passage	0° 19.056' S, 130° 55.797' E
10	30/3/01	Pulau Dua, Wai Reefs	0° 41.435' S, 130° 42.705' E
11	30/3/01	N Wruwarez I., Batanta	0° 45.448' S, 130° 46.260' E
12	30/3/01	SW Wruwarez I., Batanta	0° 47.103' S, 130° 45.865' E
13	31/3/01	Kri Island dive camp	0° 33.457' S, 130° 40.604' E
14	31/3/01	Sardine Reef	0° 32.190' S, 130° 42.934' E
15	1/4/01	Near Dayang I., Batanta	0° 47.916' S, 130° 30.274' E
16	1/4/01	NW end Batanta Island	0° 47.914' S, 130° 29.277' E
17	1/4/01	W end of Wai Reef complex	0° 42.212' S, 130° 38.847' E
18	2/4/01	Melissa's Garden, N. Fam I.	0° 35.390' S, 130° 18.909' E
19	2/4/01	N Fam Island Lagoon	0° 34.202' S, 130° 16.358' E
20	2/4/01	N tip of N Fam Island	0° 32.755' S, 130° 15.007' E
21	3/4/01	Mike's Reef, SE Gam I.	0° 31.032' S, 130° 40.304' E
22	3/4/01	Chicken Reef	0° 27.939' S, 130° 41.931' E
23	3/4/01	Besir Bay, Gam Island	0° 39.005' S, 130° 34.724' E
24	4/4/01	Ambabee I., S Fam Group	0° 44.723' S, 130° 16.547' E
25	4/4/01	SE of Miosba I., S Fam Gp.	0° 35.246' S, 130° 15.338' E
26	4/4/01	Keruo Island, N Fam Group	0° 15.741' S, 130° 18.105' E
27	5/4/01	Bay on SW Waigeo Island	0° 08.328' S, 130° 23.196' E
28	5/4/01	Between Waigeo & Kawe Is.	0° 11.924' S, 130° 07.506' E
29	5/4/01	Alyui Bay, W Waigeo	0° 01.003' N, 130° 19.690' E
30	6/4/01	N end Kawe Island	0° 00.214' N, 130° 07.904' E
31	6/4/01	Equator Islands – E side	0° 00.102' S, 130° 10.648' E
32	6/4/01	Equator Islands – W side	0° 36.815' S, 130° 09.805' E
33	7/4/01	Alyui Bay entrance, Waigeo	0° 09.912' S, 130° 13.765' E
34	7/4/01	Alyui Bay entrance, Waigeo	0° 08.942' S, 130° 13.626' E
35	7/4/01	Saripa Bay, Waigeo Island	0° 07.002' S, 130° 21.866' E
36	8/4/01	Wayag Islands – E side	0° 10.202' N, 130° 03.997' E
37	8/4/01	Wayag Islands – W side	0° 10.310' N, 130° 00.591' E
38	8/4/01	Wayag Islands – inner lagoon	0° 10.225' N, 130° 01.827' E
39	9/4/01	Ju Island, Batang Pele Group	0° 18.951' S, 130° 08.028' E
40	9/4/01	Batang Pele Island	0° 17.812' S, 130° 12.329' E
41	9/4/01	Tamagui I., Batang Pele Group	0° 19.373' S, 130° 14.720' E
42	10/4/01	Wofah Island, off SW Waigeo	0° 15.259' S, 130° 17.564' E
43	10/4/01	Between Fwoyo & Yefnab Kecil Is.	0° 24.359' S, 130° 16.200' E
44	10/4/01	Yeben Kecil Island	0° 29.256' S, 130° 20.329' E

Concurrently, the Fisheries and Reef Condition Team used a 50 m line transect placed on top of the reef to record substrate details and approximate biomass of commercially important (target) species, as well as observations on key indicator species (for assessing fishing pressure) such as groupers and Napoleon Wrasse. Additional information about utilization of marine resources was obtained through informal interviews with villagers.

The expedition used two small motor boats that served as diving platforms and rapid transport between sites. An additional motorized canoe was utilized to transport the community liaison team. Irian Diving's base camp at Kri Island was used for the initial and latter portions of the survey (27 March–3 April and 9–10 April), and the pearl farm operated by P.T. Cendana Indopearls at Alyui Bay, West Waigeo, served as our base between 4–8 April.

Details for individual sites are provided in the reef condition section (Technical Paper 5 in this report). Table 1 provides a summary of sites. Their location is also indicated on the accompanying map.

Results

Biological Diversity

The results of the RAP survey indicate an extraordinary marine fauna. Totals for the three major indicator groups (Table 2) surpassed those for previous RAPs in Indonesia, Papua New Guinea, and the Philippines.

Table 2. Summary of Raja Ampat Islands fauna recorded during the RAP survey.

Faunal Group	No. Families	No. Genera	No. Species
Reef corals	19	77	456
Molluscs	94	242	699
Fishes	93	323	970

Detailed results are given in the separate chapters for corals, molluscs, fishes, reef fisheries, reef condition, and community use of marine resources, but the key findings of the survey are summarized here.

Corals – The islands have the highest known diversity of reef corals for an area of its size. A total of 456 species plus up to nine potential new species or unusual growth forms were recorded. A remarkable 96% (565 of a total of 590) of all Scleractinia recorded from Indonesia are likely to occur in the Raja Ampat Islands.

An average of 87 species per site was recorded with the four most diverse sites as follows: Ju Island, Batang Pele Group (Site 39, 123 species), Wofah Island, off

SW Waigeo (Site 42, 122 species), Kri Island (Site 13, 115 species), and Alyui Bay (Site 29, 98 species). Relatively exposed fringing reefs supported the highest number of coral species with an average of 86.3 species per site, compared to isolated platform reefs (79.7 species) and sheltered bays (66.8). In terms of geographic areas, the reefs of the Batanta-Wai region had the highest number of corals per site (87.3) and those near the entrance of Mayalibit Bay had the lowest number (34.5).

Molluscs – Mollusc diversity was higher than for any previous RAP expedition and similar surveys conducted by the Western Australian Museum in Australia. A total of 699 species were recorded during the survey including 530 gastropods, 159 bivalves, 3 chitons, 5 cephalopods, and 2 scaphopods. The fauna was typical of that found on relatively sheltered reefs, with species associated with more exposed oceanic conditions either scarce or absent. The most diverse families were gastropods: Conidae (54 species); Muricidae (49); Cypraeidae (44); Mitridae (33); and Terebridae (28). Veneridae (28) was the most diverse bivalve family. The two richest sites for mollusc diversity were Southwest Waigeo Island (Site 27) and the Wayag Islands (Site 38) with 110 and 109 species respectively; site 36 at the eastern Wayag Islands had the lowest diversity (36 species). A number of commercially important molluscs (e.g. *Tridacna* and *Strombus*) occurred widely at the surveyed sites, but populations were invariably small.

Reef fishes – A total of 828 species were recorded, raising the total known from the islands to 970 species. An extrapolation technique utilizing six key indicator families reveals that at least 1,084 species can be expected to occur in the area. Several notable results were achieved for fishes including the two highest counts (283 and 281 species) ever recorded by G. Allen during a single dive anywhere in the world. These totals were achieved at Cape Kri (Site 2a) and at the Southern Fam Group (Site 25). A total of 200 or more species per site is considered the benchmark for an excellent fish count. These figures was achieved at 51% of Raja Ampat Sites, surpassing the previous high of 42% of sites at Milne Bay, Papua New Guinea, and well in excess of the figures for previous RAPs at the Togean-Banggai Islands (16.0%) and the Calamianes Islands, Philippines (10.5%). The average number of fish species per site was 183.6. Relatively exposed fringing reefs supported the highest number of species with an average of 208.5 species per site. Platform reefs were nearly as rich with 200.3 species per site. Sheltered Bays were relatively poor for fishes (120 per site) owing to their poor diversity of micro-habitats. Geographically the richest areas for fishes were the N. Batanta-Wai reefs (211.7 species per site) and the Fam Group (203.0 per site).

Fisheries – A total of 196 species, representing 59 genera and 19 families, were classified as target species for reef fisheries. Stocks of edible reef fishes such as fusiliers (Casionidae), snappers (Lutjanidae), jacks (Carangidae), and sweetlips (Haemulidae) were generally abundant. Similar to the situation at nearly all other reef areas in Indonesia, there was a scarcity of large groupers (Serranidae), Napoleon Wrasse (*Cheilinus undulatus*), and sharks. Although fishing pressure was judged to be light on most reefs, there is good evidence that Napoleon Wrasse and groupers are being targeted by illegal fishers for the lucrative live fish trade. Napoleon Wrasse, perhaps the paramount target of the restaurant fish trade, was seen at only seven sites (usually one fish per site). The mean total biomass estimate for sites in the Raja Ampat Islands was considerably greater than for other previously sampled areas in the "coral triangle" including Milne Bay Province (Papua New Guinea), Togean-Banggai Islands (Indonesia), and Calamianes Islands (Philippines).

Reef Condition

Reefs were generally in very good condition compared to most areas of Indonesia with high live coral diversity and minimal stress due to natural phenomenon such as cyclones, predation (i.e. crown-of-thorns starfish), and freshwater runoff. The relatively small human population exerts only light fishing pressure, and other human-induced threats appear to be minimal. Nevertheless, use of explosives for fishing, is a disturbing trend that appears to be increasing. Almost every village in the area complained about dynamite fishing by outsiders (i.e. non-villagers) and explosive damage was noted at 13.3% of the survey sites. Illegal logging (in designated nature reserve areas) and consequent siltation is also a concern. The survey team found evidence of this activity at two sites on western Waigeo.

Community Issues

A random selection of villages was visited by RAP team members, who conducted informal interviews, primarily to acquire information on the relationship between marine biodiversity and the general community. An attempt was made to assess the importance of marine resources to the economic livelihood and general well-being of local villagers. Poverty seemed to be the main concern of average villagers, and particularly the effect of the continuing economic crisis, which appears to be severely impacting their livelihood. Increased prices for basic goods such as rice and medicine are of major concern. This problem has caused increased reliance on marine products for sustenance and capital, and in some cases fishers have adopted illegal methods such as dynamite and cyanide to provide extra income to keep pace with inflation.

Outstanding Sites

The Raja Ampat Islands is generally an outstanding area for beautiful scenery and rich coral reef diversity. However, the RAP survey identified a number of sites that deserve special mention:

Cape Kri, Kri Island (Site 2a-b) – Lying off the eastern end of Kri Island, this location supports a diverse reef biota due to its incorporation of several major habitats including an exposed steep drop off (to 45 m depth), algal ridge, reef flat, and sheltered lagoon (to 28 m depth). A total of 283 fish species were noted at Site 2a, the most fishes recorded on a single dive by G. Allen from anywhere in the Indo-Pacific. The lagoon has a rich coral fauna with extensive growth of *Acropora* and foliose species. The site is further characterized by a coconut-palm beach with scenic forested hills rising steeply from the coast.

Gam-Waigeo Passage (Site 5) - This narrow, sinuous channel extends for about one kilometre and separates the islands of Gam and Waigeo. It is bounded by highly eroded, steep limestone cliffs and richly forested hills that provide a spectacular backdrop to a remarkable diving experience. There is invariably a brisk current, which divers can "ride" to their advantage. A RAP team member described the experience as "travelling down a transparent river studded with coral, sponges, gorgonians, with a profusion of brightly-colored fishes." This is the same channel that Alfred Russel Wallace sailed through when he entered Kabui Bay on 1 July 1860 (see remarks under Historical Notes).

Mayalibit Passage and adjacent Mayalibit Bay (Sites 8-9) – Although relatively poor for marine biodiversity, this is one of the most scenic attractions in the area. A narrow, winding channel bordered by high forested peaks leads from the open sea to the almost land-locked Mayalabit Bay, a 38 km-long expanse of sheltered sea that separates the western and eastern halves of Waigeo Island. The channel supports limited reef development due to extremely strong currents, but harbors an abundance of sharks. Mangroves thrive in close proximity to shoreline reefs in some sections (e.g. Site 9), and there is an abundance of mushroom-shaped islets in the southern extremity of the Bay (e.g. Site 8).

Fam Islands (Sites 18-20 and 24-26) – The Fam Group occupies about 234 km^2 and is situated at the western entrance to Dampier Strait, between Waigeo and Batanta. The archipelago consists of two hilly islands, North Fam (16 km in length) and South Fam (6 km in length with an elevation of 138 m), and a host of small rocky islets and low cays. Its waters were the clearest encountered in the Raja Ampats and harbor a wealth of marine life. The lagoon on North Fam (Site 19) and submerged reef at Site 25 were extraordinarily rich for corals and fishes respectively.

Equator Islands (Sites 31-32) – This group of small rocky islets lies off the northeastern side of Kawe Islands. It lacks a name on marine charts, therefore we dubbed it the Equator Islands because of its geographic position. The area is extremely scenic with an abundance of marine organisms. Prominent features include an expansive shallow lagoon on the eastern side and spectacular limestone "mushrooms" on the western side.

Saripa Bay (Site 35) – Saripa Bay forms a scenic, six-kilometer-long embayment near the northwestern tip of Waigeo Island. Unlike other highly sheltered bays in the area, it receives excellent tidal flushing and consequently supports a lush growth of corals and other sessile inverte-brates. Its waters were also remarkably clear compared to similar habitats.

Wayag Islands (Sites 36-38) – The Wayag Islands lie approximately 35 km northwest of the north western tip of Waigeo, and were certainly one of the highlights of the entire survey. The group consists of a veritable maze of large forested islands, tiny limestone "mushrooms," and sizeable domes or beehive-shaped islets. Marine habitats include an excellent mix of mangroves, exposed drop offs, and highly sheltered reefs, resulting in a rich marine fauna. It was the unanimous opinion of team members that the area is probably deserving of World Heritage status. It enjoys wonderful natural protection due to its remote location and lack of human inhabitants.

Conservation Recommendations

The Marine RAP survey confirms that the Raja Ampat Islands support a rich and varied marine fauna. The area is largely unspoiled, but there are disturbing signs of habitat destruction and over-exploitation of certain resources. Although much of the area is already gazetted as a wildlife reserve, there is no real enforcement of conservation laws. There is an urgent need for marine and terrestrial conserva-tion action. Unlike many other parts of Indonesia where reefs are in bad condition and overfishing is rampant, the Raja Ampats hold real promise for successful conservation due to its small human population, spectacular scenery, and extraordinary diversity of marine life. As a result of our survey we make the following specific recommendations:

1. **Implement an environmental awareness campaign.** This activity has proved successful in many other areas and is critical to any conservation program that depends on the cooperation of local communities, which is definitely the case at the Raja Ampat Islands. Local residents need to become aware of the uniqueness of their special wildlife, the wildlife's dependence on particular natural habitats, the advantages of conservation, and the consequences if no action is taken. This can be achieved in a variety of ways including primary and secondary school curricula, guest speakers at town meetings, posters, videos, etc. One of the current problems in the Raja Ampats is that many people involved with destructive conservation practices such as dynamite fishing are unaware of the long-term consequences of their actions. We need to inform villagers of the consequences and also change their attitudes toward people who are depriving the community by engaging in these practices.

2. **Promote community participation in conservation planning and management.** Local communities have a wonderful opportunity to implement and manage conservation initiatives that will play a critical role in maintaining marine biodiversity in surrounding waters. Communities need to work together to achieve the common goal of long-term reef conservation. One way of achieving this is to establish an effective team of trusted and respected elders. It is understood that Max Ammer of Irian Diving has already taken steps to organize this group. There appears to be considerable interest among local communities and the plan has been endorsed by the local Camat (regent), the highest-ranking government official in the area. One of the principle tasks of the NGO would be to achieve an effective working relationship between local communities and to resolve conflicts related to fishing rights. Essential tasks of this group would also include the arbitration of conflicts relating to marine resource utilization, and prosecution of illegal-fishing offenders. There is also a need to promote traditional knowledge of natural resource utilization and conservation.

3. **Establishment of community outreach programs to provide extra incentives for participation in conser-vation activities.** Community participation could be supported by establishing and helping to finance (via government, private corporations, and NGOs) various outreach programs that involve educational assistance, health care, and church improvements. This type of aid would certainly raise awareness of conservation and provide real rewards for villages that implement reef management programs.

4. **Establish programs to develop sustainable economic alternatives to replace illegal fishing.** If villagers are denied an income from fishing there must be sustainable alternatives to earn cash. Possible activities include eco-tourism and related activities. The Raja Ampats has great potential for tourism. Members of the RAP survey team were unanimous in their praise of the area's rich biodiversity, good reef condition, and superb physical beauty. However, if steps are not taken to halt illegal fishing practices, especially the use of explosives and cyanide, the environment will be seriously damaged. There are disturbing signs already and reliable sources indicate that destructive activities have greatly increased over the past two years. Limited development of ecotourism is an excellent method to implement conservation at the local level. Communities can reap financial benefits through tourism-related employment and also play an active part in conserving reefs. Maintaining reefs in good condition will continue to draw tourists and local communities will naturally benefit if the reefs continue to sustain their needs for marine resources. Alternatives for destructive fishing activities appear to be the only real option to protect the area. Villages could be assisted in setting up a variety of eco-tourist activities such as bird watching, hiking, kayak tours, guest houses, etc. Local guides would need to be properly trained so they are able to fully understand the expectations and demands of guests, including familiarization with scheduling procedures, booking, and public relations. Activities could also be coordinated with the research field station (see recommendation 8 below), once it was established, and also with Sorong-based agencies as the mainland offers additional tourist possibilities.

5. **Develop terrestrial and marine conservation initiatives concurrently.** The Raja Ampat Islands afford a rare opportunity to develop both terrestrial and marine conservation programs. Land and sea ecosystems are closely linked in this area and terrestrial impacts have direct consequences on marine habitats. Freshwater surveys in the Raja Ampats indicate a diverse insect and fish fauna that includes a significant endemic element. The same applies to the bird fauna, although there is still incomplete knowledge. At least two species of bird-of-paradise (the Red and Wilson's) are unique to the area. There is an urgent need for one or more terrestrial RAPs. Conservation planning for the Raja Ampats should involve careful consideration of both terrestrial and marine ecosystems.

6. **Review boundaries of existing wildlife reserves.** Current boundaries need to be reviewed to insure they can be justified and are effective for protecting a representative cross-section of all major marine and terrestrial habitats. There may be grounds for altering the boundaries or adding additional reserves after preliminary biodiversity surveys are completed. Every effort should be made to convert so called "paper parks" to meaningful reserves that are properly managed and patrolled by resident rangers.

7. **Control or eliminate illegal activities that negatively impact natural ecosystems.** A variety of threats to coastal environments originate from land-based sources. During the Raja Ampats survey we noted at least two illegal logging operations on western Waigeo Island. Uncontrolled logging not only depletes valuable natural resources, but the erosion of logged sites contributes to harmful silt deposition that directly affects coral reefs. Terrestrial RAP surveys should be undertaken in order to properly evaluate the extent of both legal and illegal logging operations on natural ecosystems of the Raja Ampat Islands. Indiscriminate destruction of natural resources and over-fishing are problems throughout Indonesia. Consequently, it may be necessary to enact more precise laws, particularly at the local level, that cover all aspects of fishing and environmental destruction.

    The overall goal of this legislation should be to sustain natural resources and conserve the natural environment for future generations. This might include laws dealing with type and quantity of gear, and catch quotas for various species based on sound biological information. Long-term catch monitoring programs would provide essential information for managing the fishery.

    Destructive fishing practices such as the use of cyanide and dynamite are illegal. However enforcement of these bans is virtually non-existent in areas such as the Raja Ampat Islands. This problem is rampant throughout Indonesia and needs to be addressed if truly effective conservation practices can be implemented. Local and national governments need to allocate funds for patrol boats, trained personnel, and other resources. Effective enforcement also needs to be backed up by adequate penalties in the form of heavy fines, confiscation of boats and fishing equipment, and/or jail sentences. Staff and equipment need to be provided to existing law enforcement agencies for the expansion of their activities into the marine environment. In practical terms there is a critical need to monitor daily fishing activities. An effective way to accomplish this would be to establish a network of manned observation posts at strategically important locations. Due to the huge size of the area, it would be best to implement this scheme over a limited area initially. Then, depending on the success of the program, it could be expanded to other areas, particularly if its initial success leads to increased government funding. One suggested priority area would include Mansuar, Gam, and North Fam Islands, which have apparently been greatly affected by recent destructive fishing practices. Each police unit would consist of a strategically located observation platform on a hill or tower, and would be equipped with living quarters, radio communication, binoculars, a small jetty, and a small, but rapid fiberglass patrol boat. Staff would be carefully selected from local villages so that all major families with fishing rights in the particular area are represented. Perhaps as many as five people could be trained as rangers for each post. A team consisting of two local rangers and one police officer would man the lookout. According to Max Ammer of Irian Diving, police officers can be hired for about US$60 per month. They would assist with observation duties and make any necessary arrests. Police presence is also necessary for protection, as offenders are known to carry weapons or frequently throw fish bombs if they are pursued.

To make sure this system is effective, the offenders should be brought to the village that owns the reef where the arrest was made. The boat would be immediately confiscated and held as bail, but the offenders would be taken to the Sorong police station.

8. **Facilitate studies that are essential for planning the conservation of marine environments.** In view of the extraordinary biodiversity of both marine and terrestrial systems, there is a need for continued in-depth studies, particularly with regards to potentially rare and endangered marine wildlife such as sharks, endemic reef organisms, dugongs, and sea turtles. Financial support of university students would provide an incentive for these studies. In addition, the establishment of a permanent biological field station with links to government agencies, Indonesian and foreign universities, and conservation NGOs is highly recommended. Such an installation would result in the long-term accumulation of biological knowledge that would be instrumental in developing sound conservation management.

9. **Promote collection of data essential for marine conservation planning.** Biological data are not the only type of information that is important for conservation planning. There is also an essential need for additional layers of geophysical, political,

ecological, cultural, and socio-economic information. All these factors need to be considered in defining a local conservation strategy. A viable option that has worked elsewhere (e.g. Milne Bay, Papua New Guinea and Togean Islands, Indonesia) would be to convene a workshop where a group of relevant experts and stakeholders review existing information to achieve consensus on a workable conservation strategy. An important result of this process would be the identification of information gaps, and proposals for how to fill them.

10. **Establish a long-term environmental monitoring program.** Periodic surveys are recommended to monitor the status of reef environments, particularly those of special significance (e.g. sites of special beauty, pristine representatives of major habitats, or places that harbor rare species). If a biological research station is established in the area, its staff and visitors could play a critical role in this regard. In addition, local communities should be involved in the monitoring process after an initial training period. Their involvement would necessitate the design of simple, yet effective, monitoring protocols that could be implemented without the presence of scientific personnel.

11. **Provide dive training for staff of local universities and conservation organizations.** There are relatively few trained divers working for NGOs and universities in Papua. Consequently, there is limited enthusiasm for marine conservation. Until now, most of the impetus has been provided by foreign NGOs. There is a genuine need for promotion of marine conservation values by Papuan biologists. One of the best ways to remedy this shortcoming is to train more people to dive, which will foster a greater appreciation for the undersea environment. The best possible spokespersons for Papua reef conservation should be native Papuan people, rather than foreigners.

12. **Conduct additional rapid assessment surveys.** There is an urgent need to conduct further marine surveys and terrestrial surveys of the Raja Ampat Islands. The area is vast and still relatively unknown biologically. The 2001 survey forms an excellent starting point, but more surveys are required. In particular, there is a need for surveys at Misool, Salawati, and eastern Waigeo, areas that were not visited during this RAP expedition. It is also important to survey terrestrial habitats to complement the marine work so that a coordinated plan for conservation in the region can be designed and implemented.

References

Allen , G. R., T. B. Werner, and S. A. McKenna (eds.). In press. A Rapid Marine Biodiversity Assessment of the coral reefs of the Togean and Banggai Islands, Sulawesi, Indonesia. Bulletin of the Rapid Assessment Program 20, Conservation International, Washington, DC.

Allen, G. R. and T. B. Werner (eds.). In press. A Rapid Marine Biodiversity Assessment of Milne Bay Province, Papua New Guinea. Second survey (2000). Bulletin of the Rapid Assessment Program, Conservation International, Washington, DC.

Gosliner, T. M, D. W. Behrens, and G. C. Williams. 1996. Coral reef animals of the Indo-Pacific. Sea Challengers, Monterey California.

Leis, J. M. 1991. Chapter 8. The pelagic stage of reef fishes: The larval biology of coral reef fishes. *In*: Sale, P.F. (ed.). The ecology of fishes on coral reefs. Academic Press, San Diego. Pp. 183–230.

Lesson, R. P. 1828. Description du noveau genre Ichthyophis et de plusierus espéces inédites ou peu connues de poissons, recueillis dans le voyage autour du monde de la Corvette "La Coquille". Mem. Soc. Nat. Hist. Paris v. 4: 397–412.

Lesson, R. P. 1830–31. Poissons. *In*: Duperrey, L. (ed.) Voyage austour du monde, …, sur la corvette de La Majesté La Coquille, pendant les années 1822, 1823, 124 et 1825, Zoologie. Zool. v. 2 (part 1): 66–238.

Quoy, J. R. C. and J. P. Gaimard. 1824. Voyage autour du monde, Enterpris par ordre du Roi exécuté sur les corvettes de S. M. "L'Uranie" et "La Physicienne" pendant les années1817, 1818, 1819, et 1820, par M. Louis de Freycinet. Zool. Poissons: 183–401.

Quoy, J. R. C. and J. P. Gaimard. 1834. Voyage de découvertes de "L'Astrolabe" exécuté par ordre du Roi, pendant les années1826-1829, sous le commandement de M. J. Dumont d'Urville. Poissons III: 647–720.

Werner, T. B. and G. R. Allen, (eds.). 2001. A Rapid Marine Biodiversity Assessment of the Calamianes Islands, Palawan Province, Philippines. Bulletin of the Rapid Assessment Program 17, Conservation International, Washington, DC.

Chapter 1

Reef corals of the Raja Ampat Islands, Papua Province, Indonesia

Part I. Overview of Scleractinia

J.E.N. Veron

Ringkasan

- Sebanyak 456 spesies dari 77 genus berhasil ditemukan selama survei RAP di Kepulauan Raja Ampat

- Selain itu, terdapat 9 spesies yang belum diketahui nama spesiesnya oleh penulis. Untuk mengetahuinya diperlukan penelitian lanjutan, dan beberapa dari jumlah tersebut kemungkinan merupakan taksa baru.

- Sebanyak 490 spesies sebelumnya telah diketahui berada di wilayah timur Indonesia dan sebanyak 581 spesies telah tercatat berada di Indonesia . Tambahan sembilan species yang belum diketahui itu menambah jumlah spesies di Indonesia menjadi 590.

- Penelitian ini menyimpulkan bahwa sebanyak 565 spesies (dari data sebelumnya, catatan terbaru dan 9 spesies yang belum diketahui) berhasil dicatat, dan atau berada di Kepulauan Raja Ampat.

- Disimpulkan bahwa 91% (565 dari total 590) Scleractinia yang ada di Indonesia dapat ditemukan, dan 79% (465 dari total 590) tercatat keberadaannya di Kepulauan Raja Ampat. Persentase ini kemungkinan akan sedikit berkurang seiring dengan penelitian lanjutan di wilayah Indonesia lainnya yang memiliki keragaman karang tinggi.

Editor's Note

Reef-building corals are reported in the following two sections. The first section by J.E.N. Veron presents an overview of scleractinian corals observed or collected during the survey (also refer to Appendix 1). The second section by D. Fenner focuses on species inventories that were compiled for individual sites (also refer to Appendix 2). The difference in total number of species (456 and 331) recorded by these authors reflects different objectives of the two studies. Fenner was chiefly concerned with designated survey sites, whereas Veron aimed to compile a species inventory of the whole area.

Summary

- A total of 456 described species and 77 genera were observed during the RAP survey at the Raja Ampat Islands.

- An additional nine species were observed that are unknown to the author. These await further study and some are possibly new taxa.

- A total of 490 species were previously recorded from far eastern Indonesia and a total of 581 species have been recorded for all of Indonesia. The addition of the nine unknown species boosts the total to 590.

- This study concludes that a total of 565 species (previous and new records plus nine unknown species) have been recorded, and/or are likely to occur, in the Raja Ampat Islands.

- It is concluded that 91% (565 of a total of 590) of all Scleractinia recorded in Indonesia *may* occur, and 79% (456 of a total of 590) *have* been recorded at the Raja Ampat Islands. This percentage would probably be slightly reduced with further study in other Indonesian regions with a high coral diversity.

Methods

Observations were recorded while utilizing scuba gear at each dive site to a maximum depth of approximately 50 m. All records are based on visual identification made underwater, except where skeletal detail was required for species determination. In the latter case, reference specimens were studied at the Australian Institute of Marine Science.

Sites were as listed elsewhere in this report except for sites 36–44, which were not visited by this author due to the necessity for an early departure.

This author's work concentrated on building a cumulative total of species for the entire island group rather than site comparisons (see Fenner, this report).

References for this work are as listed in Veron (2000). Geographic information is derived from a GIS database from which the maps in Veron (2000) were derived.

Specimens of *Porites* and *Montipora* were collected for molecular studies, and results are not yet available.

Results

Results are presented in Appendix 1. They include visual records made during the field work together with voucher specimens that were studied at the Australian Institute of Marine Science. They do not include nine additional species belonging to *Goniopora, Acropora, Anacropora,* and *Montipora* that require further study. Other genera, notably *Porites*, also warrant additional study.

McKenna et al. provide general site information elsewhere in this report. Three sites were found that exhibited extraordinary coral diversity: Melissa's Garden, Fam Islands (Site 18), North Fam Island Lagoon (Site 19), and Saripa Bay, Waigeo Island (Site 35). Of these, the first was well surveyed during this study, but the second and third warrant much more detailed work and are likely to contain species not included in this report.

Discussion

Biogeographic context of Raja Ampat Islands

Note: the total numbers of species indicated for any given region in Fig. 1 will likely be larger than that actually recorded during any field study, including the present study. Reasons for this are that the GIS database from which this map was generated.

1. Include range extrapolations which are justified on biogeographic grounds but which may not occur in reality and

2. Effectively assume that all habitat types have been surveyed. In this study, very high diversity was recorded in only a small (<5) number of sites. Thus the total species diversity recorded was site dependent, as is normal for all such studies.

In the present study, 95 of a total of 454 species records (21%) were outside the aforementioned species ranges recorded in Veron (2000) and represent range extensions. In the present study also, 60 of a total of 490 species records (12%) were predicted to occur at the Raja Ampat Islands but were not found.

This study clearly indicates that the Raja Ampat Islands are part of the global center of biodiversity, which encompasses the Indonesian-Philippines Archipelago.

Conservation merit of Raja Ampat Islands
The province as a whole has:

a. A very high proportion of all the corals of the Indonesia-Philippines archipelago (in fact the highest diversity ever recorded).

b. A majority of sites that are in good condition compared to the majority of regions in Indonesia and the Philippines, with very little damage from explosive fishing and other human impacts.

c. An extraordinary level of both underwater and terrestrial attractiveness.

These observations apply in part or in whole to other areas of Indonesia and the Philippines. However, this cannot be said of any other very large area outside the reef region from eastern Sulawesi in the west to the Papua Province in the east.

The Raja Ampat Islands can now be recognised as being part of the centre of coral biodiversity, thus an integral part of the Coral Triangle. Based on its high level of diversity, overall reef condition, and general attractiveness, the area has extremely good potential for marine conservation.

References

Veron J. E. N. 2000. Corals of the World Vols.1–3. Australian Institute of Marine Science.

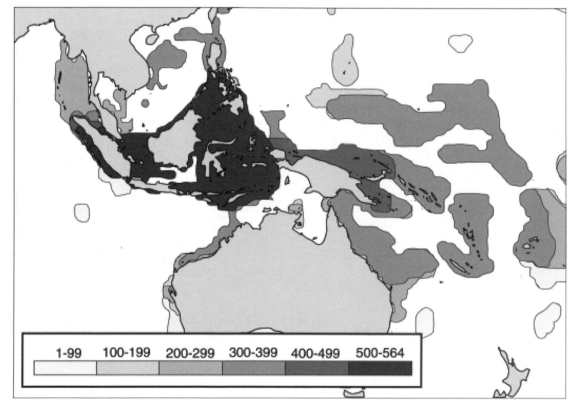

Figure 1 shows contours of diversity of zooxanthellate Scleractinia based on the GIS database of Veron. The map concentrates on areas most relevant to this report.

Reef corals of the Raja Ampat Islands, Papua Province, Indonesia

Part II. Comparison of Individual Survey Sites

Douglas Fenner

Ringkasan

- Daftar spesies karang diperoleh dari 45 lokasi. Survei ini menghabiskan waktu 51 jam penyelaman oleh D. Fenner pada kedalaman maksimum 34 meter.

- Berhasil ditemukan sebanyak 331 spesies karang batu, mewakili 76 genus dan 19 famili. (294 spesies, 67 genus dan 15 famili merupakan zooxanthella Scleractinia.

- Jumlah spesies bervariasi dari 18 sampai 123, dengan rata-rata 87 per lokasi

- *Acropora*, *Montipora* dan *Porites* adalah genus-genus dominan, dengan masing-masing 64, 30 dan 13 spesies. Komposisi ini merupakan ciri khas/tipikal terumbu karang di Indo-Pasifik, namun jumlah spesies *Acropora* adalah yang tertinggi dari yang pernah dilaporkan di lokasi manapun.

- Mayoritas hewan karang terbanyak (95%) adalah Zooxanthella Scleractinia, dengan hanya sedikit spesies non- sclearctinia dan azooxanthella, yang juga tipikal bagi karang-karang di Indo Pasifik.

Summary

- A list of corals was compiled for 45 sites. The survey involved 51 hours of scuba diving by D. Fenner to a maximum depth of 34 meters.

- A total of 331 species of stony corals in 76 genera and 19 families (294 species, 67 genera, and 15 families of zooxanthellate Scleractinia) were recorded.

- Species numbers ranged from 18 to 123, with an average of 87 per survey site.

- *Acropora*, *Montipora*, and *Porites* were dominant genera, with 64, 30, and 13 species, respectively. This composition is typical of Indo-Pacific reefs, although the number of *Acropora* species is among the highest reported from any locality.

- The overwhelming majority (95%) of corals were zooxanthellate Scleractinia, with only a few non-scleractinian and azooxanthellate species, which is typical of Indo-Pacific reefs.

Introduction

The principle aim of the coral survey was to provide an inventory of reef-associated species (including those found on sand or other sediments within and around reefs) with special emphasis on comparisons of the fauna at the various RAP sites. The primary group of corals is the zooxanthellate Scleractinian corals. Also included are a small number of zooxanthellate non-scleractinian corals that produce large skeletons, (e.g., *Millepora, Heliopora,* and *Tubipora*: fire coral, blue coral, and organ-pipe coral, respectively), and a small number of both azooxanthellate scleractinian corals (*Tubastrea, Dendrophyllia* and *Rhizopsammia*), and azooxanthellate non-scleractinian corals (*Distichopora* and *Stylaster*).

The survey results facilitate a faunal richness comparison of the Raja Ampat Islands with other parts of Southeast Asia and adjoining regions. However, the coral species list presented below is still incomplete, due to the time restriction of the survey (15 days), the highly patchy distribution of corals, and the difficulty in identifying some species underwater.

Methods

Corals were surveyed during 51 hours of scuba diving at 45 sites by D. Fenner to a maximum depth of 34 m. A list of coral species was recorded at 45 of these sites. The basic method consisted of underwater observations, usually during a single, 70 minute dive at each site. The name of each identified species was indicated on a plastic sheet on which species names were preprinted. A direct descent was made in most cases to the reef base to or beyond the deepest coral visible. The bulk of the dive consisted of a slow ascent along a zigzag path to the shallowest reef point or until further swimming was not possible. Sample areas of all habitats encountered were surveyed, including sandy areas, walls, overhangs, slopes, and shallow reef. Areas typically hosting few or no corals, such as grass beds, were not surveyed. Many corals can be confidently identified to species level while diving, but others require microscopic examination of skeletal features. References used to aid the identification process included Best and Suharsono (1991), Boschma (1959), Cairns and Zibrowius (1997), Claereboudt (1990), Dai (1989), Dai and Lin (1992), Dineson (1980), Fenner (in press), Hodgson 1985; Hodgson and Ross, 1981; Hoeksema, 1989; Hoeksema and Best (1991), Hoeksema and Best (1992), Moll and Best (1984), Nemenzo (1986), Nishihira (1986), Ogawa and Takamashi (1993 and 1995), Randall and Cheng (1984), Sheppard and Sheppard (1991), Suharsono

(1996), Veron (1985, 1986, 1990, and 2000), Veron and Nishihira (1995), Veron and Pichon, (1976, 1980, and 1982), Veron, Pichon, and Wijman-Best (1977), Wallace (1994, 1997a, and 1999), and Wallace and Wolstenholme (1998).

Results

A total of 331 species and 76 genera of stony corals (294 species and 67 genera of zooxanthellate Scleractinia) were recorded during the survey (Appendix 2). All species are illustrated in Veron (2000). The total coral species for the Raja Ampat Islands slightly surpasses the totals resulting from previous RAP surveys, which used the same methodology: 303 species at the Calamianes Islands, Philippines, 315 species at the Banggai-Togean Islands, off central Sulawesi, and 318 species at Milne Bay, Papua New Guinea.

General faunal composition

The coral fauna consists mainly of Scleractinia. The genera with the largest numbers of species include *Acropora, Montipora, Porites, Fungia, Pavona, Leptoseris, Psammocora, Astreopora, Echinopora, Favia, Echinophyllia,* and *Platygyra*. These 10 genera accounted for about 54% of the total observed species (Table 1).

Table 1. Most speciose genera of Raja Ampat corals.

Genus	Species
Acropora	68
Montipora	30
Porites	13
Fungia	11
Pavona	11
Leptoseris	8
Psammocora	7
Astreopora	6
Echinopora	6
Favia	6
Echinophyllia	6
Platygyra	6

The dominant genera are typical for western Pacific reefs (Table 2). Although their order is variable, *Acropora, Montipora,* and *Porites* are generally the three most speciose genera. The farther down the list one moves, the more variable the order becomes, with both the number of species and the differences between genera decreasing.

Table 2. Dominant genera at various western Pacific localities including eastern Australia (EA), western Australia (WA), Philippine Islands (PI), Japan Archipelago (JA), Calamianes Islands, Philippines (CI), Banggai-Togean Islands, Indonesia (BT), Milne Bay Province, Papua New Guinea (MB), and the Raja Ampat Islands (RA). Data for Australia, Philippines and Japan are from Veron (1993) and the remainder represent results of various CI RAP surveys.

Percentage of Fauna

	EA	WA	PI	JA	CI	BT	MB	RA
Acropora	19	18	17	19	13	16	22	21
Montipora	9	10	10	9	7	5	10	9
Porites	5	4	6	6	3	3	6	4
Favia	4	4	4	4	3	2	4	2
Goniopora	4	4	3	4	1	1	3	1
Fungia	4	3	4	3	4	3	4	3
Pavona	2	3	3	3	4	2	3	3
Leptoseris	2	2	2	3	4	2	3	2
Cycloseris	3	2	3	2	1	1	1	2
Psammocora	2	3	2	2	2	2	2	2

About 95% of the species were zooxanthellate (algae-containing, reef-building) scleractinian corals. The remaining corals were either azooxanthellate Scleractinia (lacking algae; four species) or non-scleractinians (12 species). A total of 72 genera and 19 families were recorded, including 64 genera and 15 families of zooanthellate scleractinans.

Zoogeographic affinities of the coral fauna

The reef corals of the Raja Ampats Islands, and Papua in general, belong to the overall Indo-West Pacific faunal province. A few species span the entire range of the province, but most do not. The Raja Ampat Islands are within the central area of greatest marine biodiversity, referred to as the Coral Triangle. The area of highest biodiversity in corals appears to be an area enclosing the Philippines, central and eastern Indonesia, northern New Guinea (Hoeksema 1992), and eastern Papua New Guinea. Areas of lower diversity include eastern Australia's Great Barrier Reef, southern New Guinea, and the Ryukyu Islands of southern Japan. Some evidence (Best, et al 1989) indicates western Indonesia may not be included in the area of highest diversity.

Reduction in species occurs in all directions away from the Coral Triangle, reaching 80 species in the Japanese Archipelago near Tokyo, 65 species at Lord Howe Island in the southwest Pacific, about 65 species in the Hawaiian Islands, and about 20 species on the Pacific coast of Panama. Species attenuation is significantly less to the west in the Indian Ocean and Red Sea. About 300 species are currently known from the Red Sea, although this area is insufficiently studied to provide accurate figures.

Corals are habitat-builders and appear to have less niche-specialization than some other groups. Some zonation occurs by depth and exposure to waves or currents. Thus, there are a few corals that are restricted to zones such as very shallow areas, protected areas, deep water, shaded niches, soft bottoms, or exposed areas. However, many corals can be found over a relatively wide range of exposure and light intensity. Corals are primarily autotrophic, relying on the products of the photosynthesis of their symbiotic algae, supplemented by plankton caught by filter-feeding and suspension feeding. Most require hard substrate for attachment, but a few grow well on soft substrates.

Table 3 presents the average number of coral species according to reef types. Fringing reefs were slightly more speciose than platform reefs, with the fewest average number of species occurring on sheltered reefs.

Table 3. Average number of species for major reef types.

Type of reef	Average no. spp.
Fringing Reefs	86
Platform Reefs	80
Sheltered Reefs	67

Coral reefs were sampled in seven different regions of the Raja Ampat islands. Coral diversity was highest in the Batang Pele to Pulau Yeben region, and lowest in the Mayalibit Bay area, as seen in Table 4.

Table 4. Average number of coral species per site recorded for geographic areas in the Raja Ampat Islands.

Area	No. sites	Average no. spp.
Batang Pele to Pulau Yeben	5	103
Batanta-Wai	6	87
Alyui Bay	6	85
Fam Islands	6	78
Gam-Mansuar	13	75
Kawe-Wayag Islands	7	69
Mayalibit Bay	2	35

The majority of coral species have a pelagic larval stage, with a minimum of a few days pelagic development for broadcast spawners (a majority of species), and larval settling competency lasting for at least a few weeks. In addition, a few species release brooded larvae that have variable dispersal capabilities ranging from immediate settlement to a lengthy pelagic period.

Most corals occurring in the Raja Ampat Islands have relatively wide distributions in the Indo-Pacific; 89% have ranges that extend both to the east and west of Indonesia, seven percent have ranges extending in two or three directions from Indonesia, but not both east and west, and five percent have ranges that extend away from Indonesia in only one direction. Only three percent of the Raja Ampat species are restricted to the coral triangle, including two that are known only from Indonesia (*Echinophyllia costata* and *Halomitra meierae*). Thus far, the Indonesian fauna includes ten possible endemics (*Acropora suharsoni*, *A. togianensis*, *A. convexa*, *A. minuta*, *A. pectinatus*, *A. parahemprichi*, *Pachyseris involuta*, *Halomitra maeierae*, *Galaxea cryptoramosa*, and *Echinophyllia costata*), of which six were described in 2000. There is a good chance that many of these recently described corals will subsequently be found in adjacent regions with further collecting activity.

Diversity at individual sites

The ten richest sites for coral diversity are indicated in Table 5. The three top sites 39, 42, and 13 had totals exceeding 100 species. The number of species recorded for every site is presented in Table 6.

Table 5. Ten richest coral sites during Raja Ampat survey.

Site No.	Location	Total spp.
39	Ju Island, Batang Pele Group	123
42	Wofah Island, off SW Waigeo	122
13	Kri Island dive camp	115
29	Alyui Bay, W Waigeo	98
7	Mios Kon Island	97
11	N Wruwarez I., Batanta	97
31	Equator Islands – E side	97
41	Tamagui I., Batang Pele Group	97
44	Yeben Kecil Island	97
2b	Cape Kri Lagoon	95

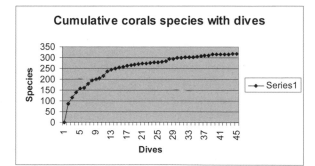

Figure 1. Coral species accumulation curve for Raja Ampats RAP survey.

Species were added to the overall list at a slow, but relatively steady rate after about 15 sites, indicating that sufficient sites were surveyed (Figure 1).

Table 6. Number of coral species observed at each site during survey of the Raja Ampat Islands.

Site	Species	Site	Species	Site	Species
1	86	15	92	30	80
2a	57	16	80	31	97
2b	95	17	81	32	70
3	64	18	91	33	80
4	48	19	41	34	62
5	18	20	92	35	83
6	94	21	71	36	55
7	97	22	85	37	85
8	39	23	57	38	49
9	30	24	84	39	123
10	85	25	83	40	106
11	97	26	79	41	97
12	89	27	66	42	122
13	115	28	50	43	90
14	93	29	98	44	97

Wallace (1997b) reported 53 species of *Acropora* from the Togean Islands, compared to 28-61 species (mean = 50.4 sp.) at four other areas of Indonesia. Further collecting at the Togeans resulted in a total of 61 species (Wallace, 1999a), the highest total for this genus recorded from a limited area. Moreover, Wallace (personal communication) added several additional species on a recent visit. Previous CI RAP surveys by the author revealed 40 species of *Acropora* in the Calamianes Islands (Philippines), 52 species at the Banggai-Togean Isands, (Indonesia), 61 species at Milne Bay (Papua New Guinea), and 68 species during the current survey. Thus, the Raja Ampat Islands have one of the richest *Acropora* faunas in the world.

Species of special interest

At least nine unidentified species between coral and were collected during the survey, including several that are potentially undescribed species. In addition, 12 recently described species (Veron, 2000) were recorded (Table 7).

Other notable species included *Montipora confusa*, *Montipora florida*, *Pachyseris foliosa*, and *Oxypora crassispinosa*, which were previously reported only from the Philippines, and were presumed to be endemics. At the Raja Ampat Islands, the first two species were common, and the latter two were uncommon. An additional two species, *Halomitra meierae* and *Echinophyllia costata*, are known only from Indonesian seas, but as both were recently described (Veron, 2000), it is likely they will eventually be collected elsewhere.

Table 7. Recently described species recorded from the Raja Ampat Islands.

Species	Previous known locality
Montipora hodgsoni	Philippines and Indonesia
Montipora palawanensis	Philippines and Papua New Guinea
Montipora verruculosus	Philippines and Papua New Guinea
Acropora cylindrica	Papua New Guinea
Porites rugosa	Sulawesi, Indonesia
Cycloseris colini	Palau Islands
Halomitra meierae	Bali, Indonesia
Acanthastrea subechinata	Philippines and Indonesia
Favia truncatus	Widespread Indo-W. Pacific
Favites paraflexuosa	Widespread Indo-W. Pacific
Platygyra acuta	Widespread Indo-W. Pacific

A total of 46 species, which have not been previously reported from Indonesia in published reports (Tomascik, 1997; Wallace, 1999), were observed or collected. Some of these were recorded during the Banggai-Togeans RAP in 1998. These species are presented in Table 8.

Overview of the Indonesian coral fauna

The Indonesian coral fauna is undoubtedly one of the richest in the world. The only other countries with comparable coral diversity are the Philippines and Papua New Guinea.

The total number of species found in this study, 330, is slightly less than that (334) reported in the most thorough single study of Indonesian corals yet reported (Best et al., 1989), and significantly more than a brief study of reefs near Jakarta (Moll & Suharsono, 1986), where 193 species were found, and a previous study near Jakarta (Brown, et al. 1985) where only 88 species were recorded (the first two studies reported only zooxanthellate Scleractinia) A recent literature review of Indonesian corals (Tomascik, 1997) reported a total of 384 species, excluding *Acropora*. Wallace (1994, 1997a, 1998, 1999) reported 94 *Acropora* from Indonesia. In addition, 46 named species were found in this study, which have not been previously reported from Indonesia in published studies (see Table 8), although 28 of these species were reported during a previous RAP in the Banggai-Togean Islands.

Including the present report, some 488 species of Scleractinia (and a total of 544 species of stony corals) have been reported from Indonesia. Additional species are undoubtedly recorded in unpublished records held by J. E. N. Veron. Eastern Indonesia, including the Raja Ampats, are clearly within the area of maximum coral diversity in the Western Pacific (although western Indonesia may be outside this area—See Best et al., 1989).

Table 8. Coral species recorded from the Raja Ampat Islands that have not been previously reported in the literature from Indonesia. Species previously known only from the Philippines (and thus thought to be endemic there), rare species, and species previously found during the 1998 RAP of the Banggai and Togean Islands are also indicated.

Species	Philippines	Rare	Banggai - Togeans
Seriatopora aculeata			
Stylophora subseriata			
Montipora altisepta			
Montipora cactus			x
Montipora capitata			x
Montipora cebuensis			
Montipora confusa			
Montipora florida	x		x
Montipora gaimardi			
Montipora mactanensis			
Montipora samarensis			
Acropora cophodactyla			
Acropora pinguis			
Acropora rosaria			
Acropora vermiculata			
Astreopora randalli			x
Astreopora suggesta			x
Porites attenuata			x
Porites evermanni			x
Porites monticulosa			x
Goniopora pendulus			x
Pavona bipartita			x
Pavona minuta (=xarife)			x
Pachyseris foliosa	x		x
Pachyseris gemmae			x
Cycloseris colini		x	
Fungia klunzingeri			x
Halomitra meierae		x	
Galaxea paucisepta			x
Pectinia maxima			
Pectinia teres			x
Hydnophora pilosa			x
Echinophyllia patula			
Mycedium mancaoi			x
Oxypora crassispinosa	x		x
Oxypora glabra			x
Blastomussa wellsi			
Acanthastrea hemprichi			
Lobophyllia robusta			x
Symphyllia hassi			x
Echinopora pacificus			x
Montastrea salebrosa			x
Euphyllia paradivisa	x		x
Euphyllia yaeyamensis			x
Stylaster sp. 1			x
Distichopora violacea			x

Corals have been extensively surveyed in the Philippines, where the most recent published count (Veron & Hodgson, 1989) of 411 Scleractinia is less than that for Indonesia, but additional unpublished data raise this total considerably. The CI RAP report for Milne Bay, concluded that about 487 Scleractinia and 500 stony corals are currently known from Papua New Guinea. It is clearly evident that Indonesia, Papua New Guinea, and the Philippines all possess rich coral faunas and the collective region represents the center of diversity. However, there is no localized area within this region of high diversity that has been completely inventoried.

Despite the incomplete nature of collections throughout the Coral Triangle, it is interesting to compare the results of the Raja Ampats survey with data from CI's previous RAP expeditions. The total number of species found at the various locations was similar: 330 species at the Raja Ampats, 311 species at the Banggai-Togeans, 304 species at the Calamianes Islands, and 318 species at Milne Bay. Similarly, 72 genera were found in the Raja Ampats, 76 at the Banggai-Togeans, 75 at the Calamianes Islands, and 77 at Milne Bay. Thus, the Raja Ampats had the highest number of species, in the lowest number of genera, of the four locations. An average of 87 species per site was recorded for the Raja Ampats compared to, 70 species for the Banggai-Togean Islands, 93 species for the Calamianes Islands and 81 species for Milne Bay.

Although each CI Marine RAP utilized the same methodology, they involved different numbers of sites. Therefore, it is possible that those with the most sites may have produced more extensive species lists. The smallest number of sites for all RAP to date is 37, for the Calamianes Islands, Philippines, where 304 species were recorded. In comparison, after 37 sites a total of 291 species was recorded at the Banggai-Togean Islands, 290 species at Milne Bay, and 311 species at the Raja Ampat Islands. These figures indicate that the Raja Ampat Islands may be just slightly more diverse than the other sites.

References

Best, M. B. and B. W. Hoeksema. 1987. New observations on Scleractinian corals from Indonesia: 1. Free-living species belonging to the Faviina. Zool. Meded. Leiden. 61: 387–403.

Best, M. B. and Suharsono. 1991. New observations on Scleractinian corals from Indonesia: 3. Species belonging to the Merulinidae with new records of *Merulina* and *Boninastrea*. Zool. Meded. Leiden. 65: 333–342.

Borel Best, M., B. W. Hoeksema, W. Moka, H. Moll, Suharsono and I. Nyoman Sutarna. 1989. Recent scleractinian coral species collected during the Snellius-II Expedition in eastern Indonesia. Netherlands J. Sea Res. 23: 107–115.

Boschma, H. 1959. Revision of the Indo-Pacific species of the genus *Distichopora*. Bijd. tot de Dier. 29: 121–171.

Brown, B.E., M. C. Holley, L. Sya'rani and M. Le Tissier. 1984. Coral diversity and cover on reef flats surrounding Pari Island, Java Sea. Atoll Res. Bull. 281: 1–17.

Cairns, S. D. and H. Zibrowius. 1997. Cnidaria Anthozoa: Azooxanthellate Scleractinia from the Philippines and Indonesian regions. In: A. Crozier and P. Bouchet (eds.), Resultats des Campagnes Musorstom, Vol 16, Mem. Mus. nat. Hist.nat. 172: 27–243.

Claereboudt, M. 1990. *Galaxea paucisepta* nom. nov. (for *G. pauciradiata*), rediscovery and redescription of a poorly known scleractinian species (Oculinidae). Galaxea. 9: 1–8.

Dai, C-F. 1989. Scleractinia of Taiwan. I. Families Astrocoeniidae and Pocilloporiidae. Acta Ocean. Taiwan. 22: 83–101.

Dai, C-F. and C-H. Lin. 1992. Scleractinia of Taiwan III. Family Agariciidae. Acta Ocean. Taiwan. 28: 80–101.

Dineson, Z. D. 1980. A revision of the coral genus *Leptoseris* (Scleractinia: Fungiina: Agariciidae). Mem. Queensland Mus. 20: 181–235.

Fenner, D. In press. Corals of Hawaii. Sea Challengers, Monterey.

Hodgson, G. 1985. A new species of *Montastrea* (Cnidaria, Scleractinia) from the Philippines. Pacific Sci. 39: 283–290.

Hodgson, G. and M. A. Ross. 1981. Unreported scleractinian corals from the Philippines. Proc. 4th Int. Coral Reef Symp. 2: 171–175.

Hoeksema, B. W. 1989. Taxonomy, phylogeny and biogeography of mushroom corals (Scleractinia: Fungiidae). Zool. Verhand. 254: 1–295.

Hoeksema, B. W. 1992. The position of northern New Guinea in the center of marine benthic diversity: a reef coral perspective. Proc. 7th Int. Coral Reef Symp. 2: 710–717.

Hoeksema, B. W. and M. B. Best. 1991. New observations on scleractinian corals from Indonesia: 2. Sipunculan-associated species belonging to the genera *Heterocyathus* and *Heteropsammia*. Zool. Meded. 65: 221–245.

Hoeksema, B. and C-F. Dai. 1992. Scleractinia of Taiwan. II. Family Fungiidae (including a new species). Bull. Inst. Zool. Acad. Sinica. 30: 201–226.

Moll, H. and M. B. Best. 1984. New scleractinian corals (Anthozoa: Scleractinia) from the Spermonde Archipelago, south Sulawesi, Indonesia. Zool. Meded. 58: 47–58.

Moll, H. and Suharsono. 1986. Distribution, diversity and abundance of reef corals in Jakarta Bay and Kepulauan Seribu. UNESCO Rep. Mar. Sci. 40: 112-125.

Nemenzo, F. Sr. 1986. Guide to Philippine Flora and Fauna: Corals. Natural Resources Management Center and the University of the Philippines, Manila.

Nishihira, M. 1991. Field Guide to Hermatypic Corals of Japan. Tokai University Press, Tokyo. (in Japanese)

Nishihira, M. and J. E. N. Veron. 1995. Corals of Japan. Kaiyusha Publishers Co., Ltd, Tokyo. (in Japanese)

Ogawa, K., and K. Takamashi. 1993. A revision of Japanese ahermatypic corals around the coastal region with guide to identification- I. Genus *Tubastraea*. Nankiseibutu: Nanki Biol. Soc. 35: 95–109. (in Japanese)

Ogawa, K. and K. Takamashi. 1995. A revision of Japanese ahermatypic corals around the coastal region with guide to identification- II. Genus *Dendrophyllia*. Nankiseibutu: Nanki Biol. Soc. 37: 15–33. (in Japanese)

Randall, R. H. and Y-M. Cheng. 1984. Recent corals of Taiwan. Part III. Shallow water Hydrozoan Corals. Acta Geol. Taiwan. 22: 35–99.

Sheppard, C. R. C. and A. L. S. Sheppard. 1991. Corals and coral communities of Arabia. Fauna Saudi Arabia 12: 3–170.

Suharsono. 1996. Jenis-jenis karang yang umum dijumpai di Perairan Indonesia. Lembaga Oseanologi Nasional (LIPI).

Tomascik, T. 1997. Classification of Scleractinia (pages 267–315). *In*: Tomascik, T., A. J. Mah, A. Nontji, and M. K. Moosa (eds.). The ecology of Indonesian seas. Periplus Editions, Singapore.

Veron, J. E. N. 1985. New scleractinia from Australian reefs. Rec. West. Aust. Mus. 12: 147–183.

Veron, J. E. N. 1986. Corals of Australia and the Indo-Pacific. Univ. Hawaii Press, Honolulu.

Veron, J. E. N. 1990. New scleractinia from Japan and other Indo-West Pacific countries. Galaxea 9: 95–173.

Veron, J. E. N. 1993. A Biogeographic Database of Hermatypic Corals. Australian Institutue of Marine Science Monograph 10: 1–433.

Veron, J. E. N. 2000. Corals of the World. Volumes 1-3. Townsville: Australian Institute of Marine Science.

Veron, J. E. N. and G. Hodgson. 1989. Annotated checklist of the hermatypic corals of the Philippines. Pacific Sci. 43: 234–287.

Veron, J. E. N. and M. Pichon. 1976. Scleractinia of Eastern Australia. I. Families Thamnasteriidae, Astrocoeniidae, Pocilloporidae. Australian Institute of Marine Science Monograph Series 1: 1–86.

Veron, J. E. N. and M. Pichon. 1980. Scleractinia of Eastern Australia. III. Families Agariciidae, Siderastreidae, Fungiidae, Oculilnidae, Merulinidae, Mussidae, Pectiniidae, Caryophyllidae, Dendrophyllidae. Australian Institute of Marine Science Monograph Series 4: 1–422.

Veron, J. E. N. and M. Pichon. 1982. Scleractinia of Eastern Australia. IV. Family Poritidae. Australian Institute of Marine Science Monograph Series 5: 1–210.

Veron, J. E. N., M. Pichon, and M. Wijsman-Best. 1977. Scleractinia of Eastern Australia. II. Families Faviidae, Trachyphyllidae. Australian Institute of Marine Science Monograph Series 3: 1–233.

Veron, J. E. N. and C. Wallace. 1984. Scleractinia of Eastern Australia. V. Family Acroporidae. Australian Institute of Marine Science Monograph Series 6:1–485.

Wallace, C. C. 1994. New species and a new species-group of the coral genus *Acropora* (Scleractinia: Astrocoeniina: Acroporidae) from Indo-Pacific locations. Invert. Tax. 8: 961–88.

Wallace, C. C. 1997a. New species of the coral genus *Acropora* and new records of recently described species from Indonesia. Zool. J. Linn. Soc. 120: 27–50.

Wallace, C. C. 1997b. The Indo-Pacific centre of coral diversity re-examined at species level. Proc. 8th Int. Coral Reef Symp. 1: 365–370.

Wallace, C. C. 1999a. The Togian Islands: coral reefs with a unique coral fauna and an hypothesized Tethys Sea signature. Coral Reefs 18: 162.

Wallace, C. C. 1999b. Staghorn corals of the world, a revision of the genus *Acropora*. CSIRO Publ., Collingwood, Australia.

Wallace, C. C. and J. Wolstenholme. 1998. Revision of the coral genus *Acropora* in Indonesia. Zool. J. Linn. Soc. 123: 199–384.

Chapter 2

Molluscs of the Raja Ampat Islands, Papua Province, Indonesia

Fred E. Wells

Ringkasan

- Laporan ini memberikan informasi tentang moluska di 44 lokasi survei di Kepulauan Raja Ampat, Propinsi Papua dari tanggal 27 Maret sampai 10 April 2001. Sebanyak mungkin habitat diteliti pada tiap lokasi untuk memperoleh daftar spesies moluska selengkapnya dalam waktu singkat.

- Ditemukan Sebanyak 699 spesies moluska: 530 Gastropoda, 159 kerang/bivalvia, 2 Scafopoda, 5 Cefalopoda dan 3 Chiton. Keragaman spesiesnya tinggi, dan menunjukkan konsistensi keragaman spesies moluska di daerah Coral Triangle yang pernah disurvey, dengan jumlah hari yang sama.

- Jumlah spesies yang ditemukan per lokasi survei berkisar 36 - 110 spesies, dengan rata-rata 74,1 ± 2,9. Keragaman spesies lebih tinggi diperoleh pada lokasi dengan tipe habitat yang lebih bervariasi.

- Kebanyakan spesies (502 atau 72%) terdapat di 5 lokasi atau kurang. Sejumlah kecil spesies (15) ditemukan di 25 lokasi atau lebih: 9 Bivalvia dan 6 Gastropoda. Beberapa spesies, seperti *Coralliophila neritoidea* dan *Tridacna squamosa* dijumpai di karang atau berasosiasi dekat dengan karang, dan spesies lainnya (*Pedum spondyloidaeum*, *Lithopaga* sp, *Arca avellana* dan *Tridacna crocea*) bersembunyi di dalam karang. *Rhinoclavis asper* hidup di areal berpasir di antara karang.

- Spesies yang paling melimpah pada setiap lokasi umumnya adalah *Pedum spondyloidaeum*, *Lithophaga* sp dan *Coralliophila neritodea*. *Tridacna crocea* melimpah di beberapa lokasi.

- Moluska yang dijumpai di Raja Ampat dikategorikan ke dalam 7 region sebaran geografi. Teluk Majalibit memiliki paling sedikit spesies (rata-rata 48,0), karena hanya 2 lokasi. Enam lokasi di Kepulauan Fam juga tergolong sedikit (rata-rata 60,1). Rata-rata jumlah spesies di 5 region geografi lainnya berkisar 71,6 spesies di Kae – Wayag sampai 86,2 spesies di Batang Pele hingga Yeben.

- Terdapat sedikit variasi keragaman moluska di tiga habitat utama terumbu karang di Kep. Raja Ampat; *sheltered bay*/teluk tersembunyi memiliki spesies paling sedikit (rata-rata 71,8); *platform reefs*/karang rata memiliki jumlah sedang (rata-rata 73,2) dan *fringing reefs*/karang tepi memiliki jumlah spesies paling banyak (76,3).

- Sebaran 258 spesies yang umum dikenal digunakan untuk menentukan pola sebaran biogeografi. Sebagian besar spesies yang diteliti, 203 (79 %) dari 258 tersebar luas di Indo-Pasifik Barat. Lima puluh empat spesies (23 %) tersebar luas di bagian barat Samudera Pasifik, bagian tengah dan barat Pasifik, atau bagian barat Pasifik dan bagian timur Samudera India. Hanya satu spesies (*Terebra caddeyi*) yang saat ini diketahui, memiliki sebaran hanya di pesisir utara New Guinea.

- Jumlah spesies moluska yang bernilai komersial ditemukan tersebar luas pada lokasi-lokasi survei. Populasinya sedikit dan kuantitas komersil tidak dijumpai. *Tridacna crocea* dan *Strombus luhuanus* melimpah pada tingkat lokal, dengan kepadatan populasi yang tidak besar.

Summary

- The present report presents information on the molluscs collected at 44 sites surveyed in Raja Ampat Islands, Papua Province, from 27 March to 10 April 2001. As many habitats as possible were examined at each site to develop as comprehensive a species list as possible of the molluscs present in the limited time available.

- A total of 699 species of molluscs were collected: 530 gastropods, 159 bivalves, 2 scaphopods, 5 cephalopods, and 3 chitons. Diversity was high, and consistent with molluscan diversity recorded on other surveys in the Coral Triangle that were undertaken over a similar number of collecting days.

- The number of species collected per site ranged from 36 to 110, with a mean of 74.1 ± 2.9. Higher diversity was recorded at sites with more variable habitat types.

- Most species (502 or 72%) occurred at five or fewer sites. A small number of species (15) were found at 25 or more sites: 9 bivalves and 6 gastropods. Some, such as *Coralliophila neritoidea* and *Tridacna squamosa*, live on or in close association with corals, and others (*Pedum spondyloidaeum*, *Lithophaga* sp., *Arca avellana*, and *Tridacna crocea*) actually burrow into coral. *Rhinoclavis asper* lives in sandy areas between the corals.

- The most abundant species at each site were generally burrowing arcid bivalves, *Pedum spondyloidaeum*, *Lithophaga* sp., and *Coralliophila neritoidea*. *Tridacna crocea* was abundant at several of the sites.

- Molluscs were categorized as occurring in seven geographical regions of the Raja Ampat Islands. Mayalibit Bay had the fewest species (mean of 48.0), but included only two sites. The six sites at the Fam Islands were also impoverished (mean = 60.1). The mean species number in five of the remaining geographical regions ranged from 71.6 at Kawe-Wayag to 86.2 at Batang Pele to Pulau Yeben.

- There was little variation between diversity of molluscs in the three major coral reef habitats in the Raja Ampat Islands: sheltered bays had the fewest species (mean of 71.8); platform reefs were intermediate (73.2); and fringing reefs had the greatest mean diversity (76.3).

- The ranges of 258 well known species were used to examine biogeographical distribution patterns. The great majority of the species studied, 203 (79%) of the 258, are widespread throughout the Indo-West Pacific. Fifty-four species (23%) are widespread in the western Pacific Ocean, the central and western Pacific, or the western Pacific and eastern Indian Oceans. Only one species (*Terebra caddeyi*) has a distribution that is presently known only from the north coast of New Guinea.

- A number of commercially important molluscs occurred widely at the surveyed sites. Populations were small, and commercial quantities were never found. *Tridacna crocea* and *Strombus luhuanus* were locally abundant, but even for these species, population densities were not great.

Introduction

In October 1997, Conservation International conducted a Marine Rapid Assessment survey of the fauna of coral reefs in Milne Bay Province, Papua New Guinea. The goal of the expedition was to collect information on the biodiversity of three key animal groups—corals, fishes and molluscs—for use in assessing the importance of the reefs for conservation purposes. Goals, methodology, and results of the expedition are described in Werner and Allen (1998); Wells (1998) described the molluscs. Following the success of the initial survey, additional surveys were conducted in the Calamianes Islands, Philippines (1998), the Togean-Banggai Islands of Indonesia (1998), and a second expedition to Milne Bay Province (2000). Molluscs are described by Wells (in press a, b). The present report describes molluscs collected or observed during the RAP survey of the Raja Ampat Islands, Papua Province, undertaken in March and April 2001.

In addition to their importance for conservation purposes, Marine Rap surveys provide an increasing dataset on biodiversity of the three target groups on reefs in the Indo-West Pacific. This complements work done in a variety of areas of the eastern Indian Ocean by the Western Australian Museum.

There appear to be no previously published reports on the marine molluscs of the Raja Ampat Islands. However, Oostingh (1925) provided a detailed report on the molluscs of the nearby Obi Major and Halmahera Islands, where 298 species were recorded. In addition, Oostingh provided a history of malacological work in the area to 1925.

Methods

The survey was conducted from 27 March to 10 April 2001, with a total of 45 sites being examined (details provided in other sections of this report). Site 2 was examined in two separate, but nearby areas, referred to in this report as 2a and 2b. All sites were surveyed by scuba diving. Each site was examined by starting at depths of 20-40 m and working up the reef slope into the shallows. Most of the time was spent in shallow (<6 m) water, as the greatest diversity of molluscs occurs in this region and diving time is also maximized. All habitats encountered at each site were examined for molluscs in order to obtain as many species as possible: living coral, the upper and lower surfaces of dead coral, shallow and deep sandy habitats, and intertidal habitats. For the same reason, no differentiation was made between species collected alive or as dead shells, as the dead shells would have been living at the site. Beach drift collections were made at a small number of sites during lunch breaks or dive intervals. These collections can significantly increase the number of species recorded at a site. Unfortunately, time available for these collections was very limited during the expedition.

This collecting approach allows the rapid assessment of species diversity over a wide variety of mollusc species. However, it is not complete. For example, no attempt was made to break open the corals to search for boring species, such as *Lithophaga*. Similarly, arcid bivalves burrowing into the corals were not thoroughly examined; nor were micro molluscs sampled. However, as the same person undertook the sampling of molluscs on all five Conservation International trips, and many of those of the Western Australian Museum, there is a good indication of relative diversity of molluscs collected on the expeditions to the various areas.

A variety of standard shell books and field guides were available for reference during the expedition. Most species were identified according to these texts, which included: Cernohorsky (1972); Springsteen and Leobrera (1986); Lamprell and Whitehead (1992); Gosliner *et al.* (1996); and Lamprell and Healy (1998).

Specimens of small species were retained in plastic vials or bags and the tissue removed with bleach. These were taken to the Western Australian Museum where they were identified using the reference collections of the Museum and specialist texts and papers on particular groups. Representatives of these species were deposited in the WA Museum. A set of reference materials of a number of the small species was also deposited in the LIPI Oseanologi collections in Jakarta.

The many distributions of the species collected have been inadequately reported in the literature, thus preventing reliable assessments of their actual occurrence. However, there are monographs or books covering a number of groups that have reliable distribution maps. The biogeographical distributions of 227 of these species were determined using the following references: Strombidae (Abbott, 1960; 1961); Terebridae (Bratcher & Cernohorsky, 1987); Phyllidiidae (Brunckhorst, 1993); Cypraeidae (Burgess, 1985); Mitridae (Cernohorsky, 1976; 1991); Nassariidae (Cernohorsky, 1984); Muricidae (Emerson & Cernohorsky, 1973; Radwin & D'Attilio, 1976; Ponder & Vokes, 1988; Houart, 1992); Cerithiidae (Houbrick, 1978; 1985; 1992); Conidae (Röckel *et al.* 1995); and Tridacnidae (Rosewater, 1965). This information was utilized to determine zoogeographic affinities of the Raja Ampats fauna.

Results

Despite the short time period available for the survey (15 collecting days), a diverse molluscan fauna was collected (Appendix 3). This consisted of a total of 699 species representing 5 molluscan classes: 530 gastropods, 159 bivalves, 2 scaphopods, 5 cephalopods, and 3 chitons (Table 1).

Table 1. Taxonomic characteristics of molluscs collected during the survey.

Class	Number of families	Number of genera	Number of species
Gastropoda	61	159	530
Bivalvia	27	76	159
Scaphopoda	1	1	2
Cephalopoda	4	4	5
Chitons	1	2	3
Total	94	242	699

The survey compares favorably with the previous Marine Rap surveys, where a range of 541 to 651 species was collected (Table 2). The Raja Ampats expedition recorded the greatest number of species (699) of any of the five Conservation International Marine RAP surveys; in particular, the number exceeded the 638–651 species collected during the two Milne Bay and Calamianes expeditions. However, the Raja Ampat expedition had only 15 collecting days in contrast to the 19 days on the first Milne Bay survey and 16 days on the Calamianes expedition. In contrast, 643 species were collected during the 11 days of the second Milne Bay expedition.

The present Raja Ampats survey also exceeds the results of similar collections, which have been made in Western Australia and nearby areas by the Western

Australian (WA) Museum. Diversity recorded during the Raja Ampats expedition was higher than all of the WA Museum surveys. The closest was 655 species collected in the Muiron Islands and eastern Exmouth Gulf, a shorter expedition of only 12 collecting days. It should be noted that the Muiron Island survey not only examined molluscs in the coral reefs of the Muiron Islands but also the extensive shallow mudflats and mangrove communities of the eastern portion of Exmouth Gulf.

Table 3 shows the total number of molluscs collected at each site ranged from 36 to 110 with a mean of 74.1 ± 2.9. Higher diversity was recorded at sites with more variable habitat types. The sites with the greatest diversity of molluscs (Table 4) were those with the greatest habitat diversity. In particular these sites had shallow sand in additional to the subtidal corals and intertidal rocks. Shallow sand is important both because of the species which live within it, and because dead shells are washed in from adjacent coral habitats and accumulate in the sand.

Table 3. Total number of mollusc species collected at each site.

Site	Number of species	Site	Number of species	Site	Number of species
1	88	15	86	30	76
2a	79	16	105	31	57
2b	88	17	48	32	72
3	59	18	77	33	77
4	44	19	56	34	74
5	78	20	74	35	64
6	66	21	101	36	36
7	88	22	76	37	85
8	57	23	45	38	109
9	39	24	44	39	106
10	79	25	57	40	86
11	95	26	53	41	63
12	90	27	110	42	75
13	93	28	66	43	87
14	56	29	73	44	100

Table 2. Numbers of mollusc species collected during previous Marine Rap surveys undertaken by Conservation International and similar surveys by the Western Australian Museum.

Location	Collecting days	Mollusc species	Reference
Raja Ampat Islands	15	665	Present survey
Togean-Banggai Islands, Indonesia	11	541	Wells, in press b
Calamianes Group, Philippines	16	651	Wells, in press a
Milne Bay, Papua New Guinea	19	638	Wells, 1998
Milne Bay, Papua New Guinea	11	643	Wells, in press c
Western Australian Museum Surveys			
Cocos (Keeling) Islands	20	380 on survey; total known fauna of 610 species	Abbott, 1950; Maes, 1967; Wells, 1994
Christmas Island (Indian Ocean)	12 plus accumulated data	313 on survey; approx. 520 total	Iredale, 1917; Wells et al., 1990; Wells and Slack-Smith, 2000
Ashmore Reef	12	433	Wells, 1993; Willan, 1993
Cartier Island	7	381	Wells, 1993
Hibernia Reef	6	294	Willan, 1993
Scott/Seringapatam Reef	8	279	Wilson, 1985; Wells and Slack-Smith, 1986
Rowley Shoals	7	260	Wells and Slack-Smith, 1986
Montebello Islands	19	633	Preston, 1914; Wells et al., 2000
Muiron Islands and Exmouth Gulf	12	655	Slack-Smith and Bryce, 1995
Bernier and Dorre Islands, Shark Bay	12	425	Slack-Smith and Bryce, 1996
Abrolhos Islands	Accumulated data	492	Wells and Bryce, 1997
Other surveys			
Chagos Islands	Accumulated data	384	Shepherd, 1984

The distributions of molluscs in the various areas of the Raja Ampats were variable, with no apparent pattern. The most diverse sites for molluscs (Table 4) and the least diverse sites (Table 5) were widely spread over the archipelago.

Table 4. Twelve richest sites for mollusc diversity in the Raja Ampat Islands.

Site	Location	Number of species
27	Bay on Southwest Waigeo Island	110
38	Wayag Islands – inner lagoon	109
39	Ju Island, Batang Pele Group	106
16	Northwest end Batanta Island	105
21	Mike's Reef, Southeast Gam I.	101
44	Yeben Kecil Island	100
11	Northern Wruwarez I., Batanta	95
13	Kri Island dive camp	93
12	Southwest Wruwarez I., Batanta	90
7	Mios Kon Island	88
1	West Mansuar Island	88
2b	Cape Kri Lagoon	88

Table 5. Eleven poorest sites for mollusc diversity in the Raja Ampat Islands.

Site	Location	Number of species
36	Wayag Islands – Eastern side	36
9	Mayalibit Passage	39
4	N Kabui Bay, West Waigeo	44
24	Ambabee Is., South Fam Group	44
23	Besir Bay, Gam Island	45
17	West end of Wai Reef complex	48
26	Keruo Island, North Fam Group	53
19	Northern Fam Island Lagoon	56
14	Sardine Reef	56
25	Southeast of Miosba Is., South Fam Group	57
31	Equator Islands – East side	57

As with the other groups examined, molluscs were categorized as occurring in seven geographical regions of the Raja Ampat Islands (Table 6). Mayalibit Bay had the fewest species, with a mean of 48.0. However, only two sites were examined in this area, so the validity of the data is questionable. The six sites at the Fam Islands were also impoverished, with a mean of 60.1 species per site. The mean species number in the five remaining geographical regions ranged from 71.6 at Kawe-Wayag to 86.2 at Batang Pele to Palau Yeben.

The mollusc data were also separated into the three main coral reef habitats occurring in the Raja Ampat Islands (Table 7). Surprisingly, there was little variation between habitats: sheltered bays had the fewest species (mean of 71.8); platform reefs were intermediate (73.2); and fringing reefs had the greatest mean diversity (76.3). The small degree of variation between the primary habitats is probably explained by the extensive small-scale variability that occurred at each dive site. Corals offer a variety of niches for molluscs: on, in or under dead and/or living corals. In addition there are sand patches between the corals that have varying components of silt. The shells of molluscs and other organisms offer sites for a number of species to live on.

A selection of molluscs for which there is adequate distributional data (as determined by recent revisions) were separated into major groups based on geographical distribution (Table 8). The great majority of the species studied, 199 (79%) of the 258, are widespread throughout the Indo-West Pacific. Fifty-four species (21%) are widespread in the western Pacific Ocean, the central and western Pacific, or the western Pacific and eastern Indian Oceans. Only one of the species (*Terebra caddeyi* Bratcher & Cernohorsky, 1982) has a restricted distribution, being presently known only from the north coast of New Guinea. However, it was only described recently and further collecting may expand its range.

Table 6. Distribution of molluscs in the seven areas covered by the survey.

Geographic area	Sites	Number of species		
		Minimum	Maximum	Mean ± 1 S.E.
Gam-Mansuar	1—7, 13, 14, 21—23	44	101	73.9 ± 5.1
Batanta-Wai	10–17	48	105	83.2 ± 8.3
Mayalibit Bay	8, 9	39	57	48.0
Fam Islands	18–20, 24–26	44	77	60.1 ± 5.3
Alyui Bay	27, 29, 33-35, 42	64	110	78.8 ± 6.5
Kawe-Wayag	28, 30, 31, 32, 36–38	36	109	71.6 ± 8.6
Batang Pele to Palau Yeben	39–44	63	106	86.2 ± 6.5
Total survey	1-44	36	110	74.1 ± 2.9

Table 7. Habitat distribution of molluscs in the Raja Ampat Islands.

Habitat	Sites	Number of species		
		Minimum	Maximum	Mean ± 1 S.E.
Fringing reefs	1, 2a, 3, 5, 6, 11, 13, 15, 16, 18, 20, 24, 26, 30-33, 36, 37, 39, 41-44	36	106	76.3 ± 3.7
Sheltered bays	2b, 4, 8, 9, 12, 19, 23, 27, 29, 35, 38, 40	39	110	71.8 ± 7.1
Platform reefs	7, 10, 14, 17, 21, 22, 25, 28, 34, 43	48	101	73.2 ± 5.2
Total survey	1-44	36	110	74.1 ± 2.9

As indicated above, a total of 45 sites were examined during the survey. Most species (502, or 72%) occurred at five or fewer sites. However, a small number of species (15) were found at 25 or more sites: nine bivalves and six gastropods (Table 9). These species can be used to characterise the dominant species on the reef. Some, such as *Coralliophila neritoidea* and *Tridacna squamosa* live on or in close association with the coral, and others (*Pedum spondyloidaeum*, *Lithophaga* sp., *Arca avellana*, and *Tridacna crocea*) actually burrow into the coral. *Rhinoclavis asper* lives in sandy areas between the corals.

The fact that these species were each found at 25 or more sites does not mean that they were all abundant, as many of the records are based on one or only a few dead shells found at the site. Many of the species (for example *Gloripallium radula*, *Antigona restriculata*, *Venus toreuma*, *Lima lima*, and *Gloripallium pallium*) were represented largely or entirely by dead shells. The most abundant species at each site were generally burrowing arcid bivalves, *Pedum spondyloidaeum*, *Lithophaga* sp., and *Coralliophila neritoidea*. *Tridacna crocea* was abundant at several of the sites.

Table 8. Geographical distribution of selected species of molluscs collected during the Raja Ampats survey.

Geographic area	Number of species	Percentage
Indo-West Pacific	203	79
Western Pacific	34	13
Central and western Pacific	12	5
Western Pacific and east Indian Ocean	8	3
Endemic to Papua New Guinea and the Coral Sea	1	0
Total	258	100

A number of commercially important mollusc species occurred widely at the surveyed sites. These include abalone (*Haliotis*), spider shells (*Lambis*), conchs (*Strombus*), *Murex ramosus*, pen shells (*Pinna* and *Atrina*), and giant clams (*Tridacna* and *Hippopus*). Populations of all of these groups were small, and commercial quantities were never found. Many species were found at only a few sites, with a high proportion of the records being only dead shells. *Tridacna crocea* and *Strombus luhuanus* were locally abundant, but in low densities.

Table 9. Most widespread species of molluscs at sites at the Raja Ampat Islands.

Species	Class	Number of sites
Arca avellana	Bivalvia	35
Gloripallium radula	Bivalvia	35
Tridacna squamosa	Bivalvia	34
Antigona restriculata	Bivalvia	32
Lithophaga sp.	Bivalvia	32
Pedum spondyloidaeum	Bivalvia	31
Coralliophila neritoidea	Gastropoda	31
Venus toreuma	Bivalvia	30
Tridacna crocea	Bivalvia	30
Gloripallium pallium	Gastropoda	26
Lima lima	Bivalvia	28
Rhinoclavis asper	Gastropoda	28
Oliva annulata	Gastropoda	26
Tectus pyramis	Gastropoda	26
Strombus microurceus	Gastropoda	25

Discussion

The survey of the Raja Ampat Islands indicates there is a diverse molluscan biota on the coral reefs of the archipelago. Clearly in a survey of only 15 days duration it is not possible to inventory all species, nor have all species at each site been recorded. However, the work demonstrates that there is excellent species diversity in the area that is well worth conserving.

There was no clear pattern of regional hotspots of molluscan diversity. Instead, the results reveal that sites with high species diversity are intermingled with those of much lower diversity. Variations occur on a small scale. Three major habitats were investigated: sheltered bays, platform reefs, and fringing reefs. There was little difference in the mean number of species in the three habitats, ranging from 67.3 in sheltered bays to 74.5 on fringing reefs.

The Raja Ampats were divided into seven geographical regions. Mayalibit Bay had the fewest species, with a mean of 47.0, but this was based on only two sites. Diversity was also low at the six sites of the Fam Islands (mean = 57.0). There was little variation in the five remaining areas, with mean diversity ranging from 67.3 at Kawe-Wayag to 81.8 at Batang Pele to Palau Yeben.

The lack of a clear pattern of high biodiversity areas suggests that marine areas of the Raja Ampats should be conserved in such a way that a representative cross section of all major habitats is included. However, more detailed surveys are necessary before a definitive plan can be considered.

Acknowledgments

I would like to warmly acknowledge the support of the other survey participants for their help in collecting specimens, exchanging ideas, and providing an enjoyable time. In addition, I very much appreciate the assistance provided by Mrs. Glad Hansen and Mr. Hugh Morrison of the Western Australian Museum in identifying specimens.

References

Abbott, R. T. 1950. Molluscan fauna of the Cocos-Keeling Islands. Bulletin of the Raffles Museum 22: 68–98.

Abbott, R. T. 1960. The genus *Strombus* in the Indo-Pacific. Indo-Pacific Mollusca 1(2): 33–144.

Abbott, R. T. 1961. The genus *Lambis* in the Indo-Pacific. Indo-Pacific Mollusca 1(3): 147–174.

Bratcher, T. and W. O. Cernohorsky. 1987. Living terebras of the world. Burlington, Massachusetts: American Malacologists.

Brunckhorst, D. J. 1993. The Systematics and phylogeny of phyllidiid nudibranchs (Doridoidea). Records of the Australian Museum, Supplement 16: 1–107.

Burgess, C. M. 1985. Cowries of the world. South Africa: Gordon Verhoef, Seacomber Publications.

Cernohorsky, W. O. 1972. Marine shells of the Pacific. Volume 2. Sydney, New South Wales: Pacific Publications.

Cernohorsky, W. O. 1976. The Mitridae of the world. Part I. The subfamily Mitrinae. Indo-Pacific Mollusca 3(17): 373–528.

Cernohorsky, W. O. 1984. Systematics of the family Nassariidae (Mollusca: Gastropoda). Bulletin of the Auckland Institute and Museum 14: 1–356.

Cernohorsky, W. O. 1991. The Mitridae of the world. Part 2. The subfamily Mitrinae concluded and subfamilies Imbricariinae and Cylindromitrinae. Monographs of Marine Mollusca 4: 1–164.

Emerson, W. K. and W. O. Cernohorsky. 1973. The genus *Drupa* in the Indo-Pacific. Indo-Pacific Mollusca 3(13): 1–40.

Gosliner, T. M., D. W. Behrens and G. C. Williams. 1996. Coral reef animals of the Indo-Pacific. Monterey, California: Sea Challengers.

Houart, R. 1992. The genus *Chicoreus* and related genera (Gastropoda: Muricidae) in the Indo-West Pacific. *Mémoires du Muséum National d'Histoire Naturelle, Zoologie* (A) 154: 1–188.

Houbrick, R. S. 1978. The family Cerithiidae in the Indo-Pacific. Part 1: The genera *Rhinoclavis, Pseudovertagus* and *Clavocerithium*. Monographs of marine Mollusca 1: 1–130.

Houbrick, R. S. 1985. The genus *Clypeomorus* Jousseaume (Cerithiidae: Prosobranchia). Smithsonian Contributions to Zoology 403: 1–131.

Houbrick, R. S. 1992. Monograph of the genus *Cerithium* Bruguière in the Indo-Pacific (Cerithiidae: Prosobranchia). Smithsonian Contributions to Zoology 510: 1–211.

Iredale, T. 1917. On some new species of marine molluscs from Christmas Island, Indian Ocean. Proceedings of the Malacological Society of London 12: 331–334.

Kay, E. A. 1979. Hawaiian marine shells. Honolulu, Hawaii: Bishop Museum Press.

Lamprell, K. and J. M. Healy. 1998. Bivalves of Australia. Volume 2. Leiden, The Netherlands: Backhuys Publishers.

Lamprell, K. and T. Whitehead. 1992. Bivalves of Australia. Volume 1. Bathurst, New South Wales: Crawford House Press.

Maes, V. O. 1967. The littoral marine molluscs of Cocos-Keeling Islands (Indian Ocean). Proceedings of the Academy of Natural Science of Philadelphia 119: 93–217.

Oostingh, C. H. 1925. Report on a collection of Recent shells from Obi and Halmahera (Moluccas). Mededeelingen van de Landbouw-Hoogeschool te Wageningen (Nederland) 29: 1–362.

Ponder, W. F. and E. H. Vokes. 1988. A revision of the Indo-West Pacific fossil and resent species of *Murex* s.s and *Haustellum* (Mollusca: Gastropoda: Muricidae). Records of the Australian Museum, Supplement 8: 1–160.

Preston, H. B. 1914. Description of new species of land and marine shells from the Montebello Islands, Western Australia. Proceedings of the Malacological Society of London 11: 13–18.

Radwin, G. E. and A. D'Attilio. 1976. Murex shells of the world. Stanford, California: Stanford University Press.

Röckel, D., W. Korn, and A. J. Kohn. 1995. Manual of the living Conidae. Volume 1. Indo-Pacific Region. Wiesbaden, Germany: Verlag Christa Hemmen.

Rosewater, J. 1965. The family Tridacnidae in the Indo-Pacific. Indo-Pacific Mollusca 1(6): 347–396.

Sheppard, A. L. S. 1984. The molluscan fauna of Chagos (Indian Ocean) and an analysis of its broad distribution patterns. Coral Reefs 3: 43–50.

Slack-Smith, S. M. and C. W. Bryce. 1995. Molluscs. *In:* Hutchins, J. B., S. M. Slack-Smith, L. M. Marsh, D. S. Jones, C. W. Bryce, M. A. Hewitt, and A. Hill. Marine biological survey of Bernier and Dorre Islands, Shark Bay. Western Australian Museum and Department of Conservation and Land Management, manuscript report. Pp. 57–81.

Slack-Smith, S. M. and C. W. Bryce. 1996. Molluscs. *In:* Hutchins, J. B., S. M. Slack-Smith, C. W. Bryce, S.M. Morrison, and M. A. Hewitt. Marine biological survey of the Muiron Islands and the eastern shore of Exmouth Gulf, Western Australia. Western Australian Museum and Department of Conservation and Land Management, manuscript report. Pp. 64–100.

Springsteen, F. J. and F. M. Leobrera. 1986. Shells of the Philippines. Manila, Philippines: Carfel Seashell Museum.

Wells, F. E. 1993. Part IV. Molluscs. *In:* Berry, P. F. (ed.) Faunal Survey of Ashmore Reef, Western Australia. Records of the Western Australian Museum, Supplement 44: 25–44.

Wells, F. E. 1994. Marine Molluscs of the Cocos (Keeling) Islands. Atoll Research Bulletin 410: 1–22.

Wells, F. E. 1998. Marine Molluscs of Milne Bay Province, Papua New Guinea. *In:* Werner, T. and Allen, G. R. (eds.). A rapid biodiversity assessment of the coral reefs of Milne Bay Province, Papua New Guinea. RAP Working Papers Number 11. Washington, DC: Conservation International. Pp. 35–38.

Wells, F. E. 2000. Molluscs of the Calamianes Group, Philippines. *In:* Werner, T.B., and G.R. Allen (eds.). A Rapid Marine Biodiversity Assessment of the Calamianes Islands, Palawan Province, Philippines. Bulletin of the Rapid Assessment Program 17, Conservation International, Washington, DC.

Wells, F. E. In press a. Molluscs of the Gulf of Tomini, Indonesia. *In:* Allen, G. R., T. B. Werner, and S. A. McKenna (eds.). A rapid biodiversity assessment of the coral reefs of the Togean and Banggai Islands,

Sulawesi, Indonesia. Bulletin of the Rapid Assessment Program, Conservation International, Washington, DC.

Wells, F. E. In press b. Molluscs of Milne Bay Province, Papua New Guinea 2000. *In:* Allen, G. R., T. Werner, and S. A. McKenna (eds.). A rapid biodiversity assessment of the coral reefs of Milne Bay Province, Papua New Guinea. Second Survey. Bulletin of the Rapid Assessment Program, Conservation International, Washington, DC.

Wells, F. E. and C. W. Bryce. 1997. A preliminary checklist of the marine macromolluscs of the Houtman Abrolhos Islands, Western Australia. *In:* Wells, F. E. (ed) Proceedings of the seventh international marine biological workshop: The marine flora and fauna of the Houtman Abrolhos Islands, Western Australia. Western Australian Museum, Perth. Pp. 362–384.

Wells, F. E., C. W. Bryce, J. E. Clarke, and G. M. Hansen. 1990. Christmas Shells: The Marine Molluscs of Christmas Island (Indian Ocean). Christmas Island: Christmas Island Natural History Association.

Wells, F. E. and S. M. Slack-Smith. 1986. Part IV. Molluscs. *In:* Berry, P. F. (ed.) Faunal Survey of the Rowley Shoals and Scott Reef, Western Australia. Records of the Western Australian Museum, Supplement 25: 41–58.

Wells, F. E. and S. M. Slack-Smith. 2000. Molluscs of Christmas Island. *In:* Berry, P. F. and F. E. Wells (eds.). Survey of the marine fauna of the Montebello Islands, Western Australia and Christmas Island, Indian Ocean. Records of the Western Australian Museum, Supplement 59: 103–116.

Wells, F. E., S. M. Slack-Smith, and C. W. Bryce. 2000. Molluscs of the Montebello Islands. *In:* Berry, P. F. and F. E. Wells (eds.). Survey of the marine fauna of the Montebello Islands, Western Australia and Christmas Island, Indian Ocean. Records of the Western Australian Museum, Supplement 59: 29–46.

Werner, T. B. and G. R. Allen (eds.). 1998. A rapid biodiversity assessment of the coral reefs of Milne Bay Province, Papua New Guinea. RAP Working Papers Number 11. Washington, DC: Conservation International.

Willan, R. C. 1993. Molluscs. *In:* Russell, B. C. and Hanley, J. R. The marine biological resources and heritage values of Cartier and Hibernia Reefs, Timor Sea. Northern Territory Museum, manuscript report.

Wilson, B. R. 1985. Notes on a brief visit to Seringapatam Atoll, North West Shelf, Australia. Atoll Research Bulletin 292: 83–100.

Chapter 3

Reef Fishes of the Raja Ampat Islands, Papua Province, Indonesia

Gerald R. Allen

Ringkasan

- Daftar spesies ikan diperoleh dari 45 lokasi di Kepulauan Raja Ampat. Survei ini menghabiskan waktu 60 jam penyeleman dengan kedalaman maksimum 46 meter.

- Di Kepulauan Raja Ampat terdapat kekayaan spesies ikan karang tertinggi di dunia, sedikitnya terdapat 970 spesies. Sebanyak 828 spesies (85%) dijumpai dan dikoleksi selama survei ini. Jumlah tersebut termasuk 4 taksa baru dari famili *Pseudochromidae*, *Apogonidae* dan *Gobiidae*.

- Rumus untuk menduga jumlah total spesies ikan karang berdasarkan jumlah spesies dari 6 famili indikator kunci, menunjukkan sekurang-kurangnya 1.084 spesies diharapkan terdapat di Kepulauan Raja Ampat.

- Ikan-ikan gobi (Gobiidae), ikan damsel (Pomacentricae), dan ikan keling/maming (Labridae) adalah kelompok dominan di Kepulauan Raja Ampat dalam jumlah spesies (secara berturut-turut 110, 109 dan 98) dan jumlah individu.

- Jumlah spesies yang ditemukan berdasarkan pengamatan langsung pada survei tahun 2001, bervariasi antara 81 sampai 283, dengan rata-rata 183,6 spesies.

- Mencatat 200 spesies atau lebih per lokasi merupakan tanda penghitungan ikan yang sangat bagus dalam suatu survei. Kondisi itu terjadi di 52% lokasi survei di Raja Ampat sedangkan di Milne Bay, Papua New Guinea hanya 42% (2000), 19% di Kepulauan Togean –Banggai, Indonesia (1998) dan 10.5% di Kepulauan Calamianes, Filipina (1998)

- Walaupun keanekaragamannya relatif tinggi, terdapat tanda-tanda adanya penangkapan ikan yang berlebihan. Napoleon/maming, yang merupakan indikator yang baik untuk mengetahui adanya tekanan, jarang ditemukan. Hanya tujuh individu yang teramati, kebanyakan berukuran kecil.

- Formasi karang di Selat Dampier yang terletak di antara bagian utara Pulau Batanta dan bagian selatan Pulau Waigeo-Gam merupakan daerah yang sangat kaya dengan spesies ikan karangnya, rata-rata 212 spesies per lokasi.

- Karang-karang tepi/*Fringing reefs* di sekitar pulau-pulau besar dan kecil memiliki keragaman spesies ikan yang sangat tinggi dengan rata-rata 208 spesies per lokasi. Habitat umum lainnya termasuk karang-karang datar/*Platform reefs* mengandung 200 spesies per lokasi dan *sheltered bay/teluk tersembunyi* 120 spesies per lokasi.

- Sejauh ini terdapat 6 spesies yang hanya dijumpai di karang-karang Kepulauan Raja Ampat; *Hemiscyllium freycineti* (Hemiscyllidae); *Pseudochromis* n. sp. (Pseudochromidae), dua spesies *Apogon* yang belum dideskripsikan (Apogonidae), *Meiacanthus crinitus* (Blennidae) dan *Eviota* n. sp (Gobiidae).

- Kawasan dengan konsentrasi keragaman ikan yang sangat tinggi sehingga perlu dilakukan upaya konservasi yaitu : Tanjung Kri dan laguna di sekitarnya, Kepulauan Fam, Kepulauan Equator dan ujung barat dari utara Pulau Batanta.

Summary

- A list of fishes was compiled for 45 sites in the Raja Ampat Islands. The survey involved 60 hours of scuba diving to a maximum depth of 46 m.

- The Raja Ampat Islands have one of the world's richest coral reef fish faunas, consisting of at least 970 species of which 828 (85%) were observed or collected during the present survey. The total includes at least four new taxa belonging to the families Pseudochromidae, Apogonidae, and Gobiidae.

- A formula for predicting the total reef fish fauna based on the number of species in six key indicator families indicates that at least 1,084 species can be expected to occur at the Raja Ampat Islands.

- Gobies (Gobiidae), damselfishes (Pomacentridae), and wrasses (Labridae) are the dominant groups at the Raja Ampat Islands in both number of species (110, 109, and 98 respectively) and number of individuals.

- Species numbers at visually sampled sites during the 2001 survey ranged from 81 to 283, with an average of 183.6.

- 200 or more species per site is considered the benchmark for an excellent fish count. This figure was achieved at 52% of Raja Ampat sites compared to 42% of sites at Milne Bay, Papua New Guinea (2000), 19% at the Togean-Banggai Islands, Indonesia (1998), and 10.5% at the Calamianes Islands, Philippines (1998).

- Although fish diversity was relatively high, there were signs of overfishing. Napoleon wrasse, which are a good indicator of fishing pressure, were relatively rare. Only seven, mainly small, individuals were observed.

- The reefs in the Dampier Strait region between northern Batanta Island and southern Waigeo-Gam Island was the richest area for reef fishes with an average of 212 species per site.

- Fringing reefs around large and small islands contained the highest fish diversity with an average of 208 species per site. Other major habitats include platform reefs (200 per site), and sheltered bays (120 per site).

- The following six species are known thus far only from reefs of the Raja Ampat Islands: *Hemiscyllium freycineti* (Hemiscyllidae); *Pseudochromis* n. sp. (Pseudochromidae), two undescribed Apogon (Apogonidae), *Meiacanthus crinitus* (Blenniidae), and Eviota n. sp. (Gobiidae).

- Areas with the highest concentration of fish diversity and consequent high conservation potential include: Cape Kri and adjacent lagoon, Fam Islands, Equator Islands, and western end of northern Batanta Island.

Introduction

The primary goal of the fish survey was to provide a comprehensive inventory of reef species inhabiting the Raja Ampat Islands. This segment of the fauna includes fishes living on or near coral reefs down to the limit of safe sport diving or approximately 50 m depth. It therefore excludes deepwater fishes, offshore pelagic species such as flyingfishes, tunas, billfishes, and most estuarine forms.

Survey results facilitate comparison of the Raja Ampat's faunal richness with adjoining regions in the Indo-Australian Archipelago ("Coral Triangle"). However, the list of Raja Ampat fishes is still incomplete, due to the rapid nature of the survey and secretive nature of many small reef species. Nevertheless, a basic knowledge of the cryptic component of the fauna in other areas, coupled with an extrapolation method utilizing key "index" families, can be used to predict the Raja Ampat's overall species total.

Methods

The fish portion of this survey involved 60 hours of scuba diving by G. Allen to a maximum depth of 46 m. A list of fishes was compiled for 45 sites. The basic method consisted of underwater observations made during a single, 60–90 minute dive at each site. The name of each observed species was recorded in pencil on a plastic sheet attached to a clipboard. The technique usually involved rapid descent to 20–46 m, then a slow, meandering ascent back to the shallows. The majority of time was spent in the 2–12 m depth zone, which consistently harbors the largest number of species. Each dive included a representative sample of all major bottom types and habitat situations, for example rocky intertidal, reef flat, steep drop-offs, caves (utilizing a flashlight if necessary), rubble and sand patches.

Only the names of fishes for which identification was absolutely certain were recorded. However, very few, less than 1% of those observed, could not be identified to species. This high level of recognition is based on more than 25 years of diving experience in the Indo-Pacific and an intimate knowledge of the reef fishes of this vast region as a result of extensive laboratory and field studies.

The visual survey was supplemented with occasional small collections procured with the use of the ichthyocide rotenone and several specimens collected with a rubber-propelled, multi-prong spear. The purpose of the rotenone collections was to flush out small crevice and sand-dwelling fishes (for example eels and tiny gobies) that are difficult to record with visual techniques.

Results

The total reef fish fauna of the Raja Ampat Islands reported herein consists of 970 species belonging to 323 genera and 93 families (Appendix 4). A total of 828 species were actually recorded during the present survey. An additional 142 species were recorded by the author during two preliminary visits in 1998–1999. Allen (1993 and 1997), Myers (1989), Kuiter (1992), and Randall et al. (1990) illustrated the majority of species currently known from the area.

General faunal composition
The fish fauna of the Raja Ampat Islands consists mainly of species associated with coral reefs. The most abundant families in terms of number of species are gobies (Gobiidae), damselfishes (Pomacentridae), wrasses (Labridae), cardinalfishes (Apogonidae), groupers (Serranidae), butterflyfishes (Chaetodontidae), surgeonfishes (Acanthuridae), blennies (Blenniidae), parrotfishes (Scaridae), and snappers (Lutjanidae). These 10 families collectively account for 61% of the total reef fauna (Fig. 1).

The relative abundance of Raja Ampat fish families is similar to other reef areas in the Indo-Pacific, although the ranking of individual families is variable as shown in Table 1. Even though Gobiidae was the leading family, it was

not adequately collected, due to the small size and cryptic habits of many species. Similarly, the moray eel family Muraenidae is consistently among the most speciose groups at other localities, and is no doubt abundant in Papua Province. However, they are best sampled with rotenone due to their cryptic habits.

Fish community structure
The composition of local reef fish communities in the Indo-Pacific region is dependent on habitat variability. The incredibly rich reef fish fauna of Indonesia directly reflects a high level of habitat diversity. Nearly every conceivable habitat situation is present from highly sheltered embayments with a large influx of freshwater to oceanic atolls and outer barrier reefs. To a certain degree, the Raja Ampat Islands present a cross-section in miniature of this impressive array of reef environments. However, due to prevailing weather conditions and the protective influence of the large islands of Waigeo and Batanta, much of the surrounding sea is inordinately calm for most of the year. Therefore, fishes usually associated with sheltered reefs are perhaps over-represented. For example, among the 108 species of pomacentrids, 34% are generally found in sheltered waters, compared with a figure of 25% for Milne Bay, Papua New Guinea, where there is significantly more exposed outer reef habitat.

Similar to other reef areas in the Indo-Pacific, most Raja Ampat fishes are benthic (or at least living near the bottom) diurnal carnivores with 79% and 62% of species being assigned to these respective categories. Approximately 10% of Raja Ampat fishes are nocturnal, 4% are cryptic crevice dwellers, 4% are diurnal mid-water swimmers, and about 3% are transient or roving predators. In addition to carnivores, the other major feeding categories include omnivores (15.1%), planktivores (14.7%), and herbivores (8.2%).

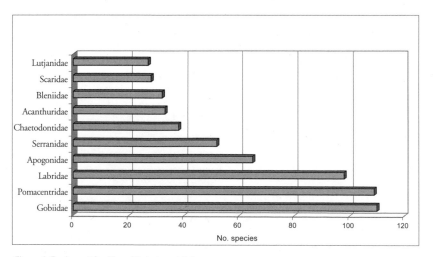

Figure 1. Ten largest families of Raja Ampat fishes

Table 1. Family ranking in terms of species number for various localities in the Indo-Pacific region. Data for Mine Bay, Papua New Guinea, is from Allen (in press), Togean-Banggai Islands, Indonesia, from Allen (in press a), for Calamianes Islands, Philippines, from Allen (2001), for the Chagos Archipelago from Winterbottom *et al.* (1989), and for the Marshall Islands from Randall and Randall (1987).

Family	Raja Ampats	Milne Bay Province	Togean-Banggai Islands	Calamianes Islands	Chagos Archipel.	Marshall Islands
Gobiidae	1st	1st	1st	3rd	1st	1st
Pomacentridae	2nd	3rd	3rd	1st	3rd	4th
Labridae	3rd	2nd	2nd	2nd	2nd	2nd
Apogonidae	4th	4th	4th	4th	6th	8th
Serranidae	5th	5th	5th	5th	4th	3rd
Chaetodontidae	6th	6th	7th	6th	11th	8th
Acanthuridae	7th	8th	8th	7th	8th	7th
Blenniidae	8th	6th	6th	8th	9th	6th
Scaridae	9th	10th	10th	10th	12th	10th
Lutjanidae	10th	9th	9th	9th	7th	18th

Table 2. Number of fish species observed at each site during survey of the Raja Ampat Islands.

Site	Species	Site	Species	Site	Species
1	220	15	219	30	210
2a	283	16	246	31	258
2b	150	17	184	32	205
3	190	18	213	33	208
4	89	19	86	34	141
5	113	20	214	35	117
6	209	21	208	36	156
7	223	22	199	37	209
8	105	23	88	38	113
9	158	24	164	39	202
10	213	25	281	40	131
11	225	26	263	41	167
12	183	27	81	42	201
13	246	28	159	43	185
14	210	29	142	44	202

Table 3. Ten richest fish sites during Raja Ampat survey.

Site No.	Location	Total fish spp.
2a	Cape Kri, Kri Island	283
25	SE of Miosba I., S Fam Is.	281
26	Keruo Island, N Fam Is.	263
31	Equator Islands – E side	258
16	NW end Batanta Island	246
13	Kri Island dive camp	244
14	Sardine Reef	226
11	N Wruwarez Is., Batanta	225
7	Mios Kon Island	223
1	W. Mansuar Island	220

The number of species found at each site is indicated in Table 2. Totals ranged from 81 to 283, with an average of 183.6 per site.

Richest sites for fishes

The total species at a particular site is ultimately dependent on the availability of food and shelter and the diversity of substrata. Well developed reefs with relatively high coral diversity and significant live coral cover were usually the richest areas for fishes, particularly if the reefs were exposed to periodic strong currents. These areas provide an abundance of shelter for fishes of all sizes and the currents are vital for supporting numerous planktivores, the smallest of which provide food for larger predators. Site 14 (Sardine Reef) is a good example of this situation. I have rarely witnessed such a dense concentration of reef fishes. Especially prominent were large shoals of *Lutjanus bohar*,

Macolor macularis, M. niger, Plectorhinchus chrysotaenia, P. polytaenia, and four species each of *Caesio* and *Pterocaesio*.

Although silty bays (often relatively rich for corals), mangroves, seagrass beds, and pure sand-rubble areas were consistently the poorest areas for fish diversity, sites that incorporate mixed substrates (in addition to live coral) usually support the most fish species. Sites that encompass both exposed outer reefs and sheltered back reefs or shoreline reefs are also correlated with higher than average fish diversity.

The 10 most speciose sites for fishes are indicated in Table 3. The average total for all sites (183.6) was high, especially considering that 12 sites were located in highly sheltered waters of deep bays, with relatively impoverished fish communities (average of 120 species per site). The total for sites 2a and 25 are the highest recorded by the author for a single dive anywhere in the Indo-Pacific.

Table 4 presents a reef fish fauna comparison of the major geographical areas that were surveyed. The highest average number of species (212) was recorded for northern Batanta and the adjacent island and reef complex in the vicinity of Wai Island. The lowest value was for the highly sheltered reefs inside Mayalibit Bay on Waigeo Island.

Table 4. Average number of fish species per site recorded for geographic areas in the Raja Ampat Islands.

Area	No. sites	Average no. spp.
Batanta-Wai	6	186.4
Fam Islands	6	203.0
Kawe-Wayag Islands	7	187.1
Gam-Mansuar	13	186.4
Batang Pele to Pulau Yeben	5	177.4
Alyui Bay	6	148.3
Mayalibit Bay	2	131.5

Coral Fish Diversity Index (CFDI)

Allen (1998) devised a convenient method for assessing and comparing overall reef fish diversity. The technique essentially involves an inventory of six key families: Chaetodontidae, Pomacanthidae, Pomacentridae, Labridae, Scaridae, and Acanthuridae. The number of species in these families is totalled to obtain the Coral Fish Diversity Index (CFDI) for a single dive site, relatively restricted geographic areas (e.g. Raja Ampat Islands) or countries and large regions (e.g. Indonesia).

CFDI values can be used to make a reasonably accurate estimate of the total coral reef fish fauna of a particular locality by means of regression formulas. The latter were obtained after analysis of 35 Indo-Pacific locations for which reliable, comprehensive species lists exist. The data were first divided into two groups: those from relatively restricted localities (surrounding seas encompassing less than 2,000 km²) and those from much larger areas (surrounding seas encompassing more than 50,000 km²). Simple regression analysis revealed a highly significant difference (P = 0.0001) between these two groups. Therefore, the data were separated and subjected to additional analysis. The Macintosh program Statview was used to perform simple linear regression analyses on each data set in order to determine a predictor formula, using CFDI as the predictor variable (x) for estimating the independent variable (y) or total coral reef fish fauna. The resultant formulae were obtained: 1. total fauna of areas with surrounding seas encompassing more than 50,000 km² = 4.234(CFDI) - 114.446 (d.f = 15; R^2 = 0.964; P = 0.0001); 2. total fauna of areas with surrounding seas encompassing less than 2,000 km² = 3.39 (CFDI) - 20.595 (d.f = 18; R^2 = 0.96; P = 0.0001).

The CFDI regression formula is particularly useful for large regions, such as Indonesia and the Philippines, where reliable totals are lacking. Moreover, the CFDI predictor value can be used to gauge the thoroughness of a particular short-term survey that is either currently in progress or already completed. For example, the CFDI for the Raja Ampat Islands now stands at 326, and the appropriate regression formula (3.39 x 326 - 20.595) predicts an approximate total of 1,084 species, indicating that approximately 114 more species can be expected.

On a much large scale the CFDI can be used to estimate the reef fish fauna of the entire Indo-west Pacific region, a frequent subject of conjecture. This method estimates a faunal total of 3,764 species, a figure that compares favorably with the approximately 3,950 total proposed by Springer (1982). Moreover, Springer's figure covers shore fishes rather than reef fishes and therefore includes species not always associated with reefs (e.g. estuarine fishes).

The total CFDI for the Raja Ampat Islands has the following components: Labridae (108), Pomacentridae (100), Chaetodontidae (42), Acanthuridae (34), Scaridae (28), and Pomacanthidae (25). Table 5 presents a ranking of Indo-Pacific areas that have been surveyed to date based on CFDI values. It also includes the number of reef fishes thus far recorded for each area, as well as the total fauna predicted by the CFDI regression formula.

The world's leading country for reef fish diversity, based on CFDI values, is Indonesia. A recent study by Allen and Adrim (in progress), which lists a total of 2,027 species from Indonesia, strongly supports this ranking. Table 6 presents CFDI values, number of shallow reef fishes recorded to date, and the estimated number of species based on CFDI data for selected countries or regions in the Indo-Pacific. In most cases the predicted number of species is similar or less than that actually recorded, and is thus indicative of the level of knowledge.

For example, when the actual number is substantially less than the estimated total (e.g. Sabah) it indicates incomplete sampling. However, the opposite trend is evident for Indonesia, with the actual number being significantly greater than what is predicted by the CFDI. The total number of species for the Philippines is yet to be determined and therefore is excluded from Table 6.

Zoogeographic affinities of the Raja Ampats fish fauna

Papua belongs to the overall Indo-west Pacific faunal community. Its reef fishes are very similar to those inhabiting other areas within this vast region, stretching eastward from East Africa and the Red Sea to the islands of Micronesia and Polynesia. Although most families and many genera and species are consistently present across the region, the species composition varies greatly according to locality.

The Raja Ampat Islands are part of the Indo-Australian region, the richest faunal province on the globe in terms of biodiversity. The nucleus of this region, or Coral Triangle, is composed of Indonesia, Philippines, and Papua New Guinea. Species richness generally declines with increased distance from the Triangle, although the rate of attenuation is generally less in a westerly direction. The damselfish family Pomacentridae is typical in this regard. For example, Indonesia has the world's highest total with 138 species, with the following totals recorded for other areas (Allen, 1991): Papua New Guinea (109), northern Australia (95), W. Thailand (60), Fiji Islands (60), Maldives (43), Red Sea (34), Society Islands (30), and Hawaiian Islands (15). The damselfishes also provide evidence that the Raja Ampat Islands are very close to the much-debated center of marine diversity. Its total of 108 species is the highest recorded for any similar-sized area in the world. Indeed, only a few countries can match this number (see discussion at end of this section).

Figure 2 presents the major zoogeographic categories for reef fishes of the Raja Ampat Islands. The largest segment of the fauna consists of species that are broadly distributed in the Indo-west and Central Pacific region from East Africa to the islands of Oceania. This is not surprising as nearly all coral reef fishes have a pelagic larval stage of variable duration, depending on the species. Dispersal capabilities and length of larval life of a given species are usually reflected in its geographic distribution. A substantial percentage of Raja Ampat fishes are confined to the species-rich Indo-Australian Archipelago. These are mainly species that seem to lack efficient dispersal capabilities and are therefore unable to exploit oceanic habitats.

Table 5. Coral fish diversity index (CFDI) values for restricted localities, number of coral reef fish species as determined by surveys to date, and estimated numbers using the CFDI regression formula (refer to text for details).

Locality	CFDI	No. reef fishes	Estim. reef fishes
Milne Bay, Papua New Guinea	337	1109	1313
Maumere Bay, Flores, Indonesia	333	1111	1107
Raja Ampat Islands, Indonesia	326	972	1084
Togean and Banggai Islands, Indonesia	308	819	1023
Komodo Islands, Indonesia	280	722	928
Madang, Papua New Guinea	257	787	850
Kimbe Bay, Papua New Guinea	254	687	840
Manado, Sulawesi, Indonesia	249	624	823
Capricorn Group, Great Barrier Reef	232	803	765
Ashmore/Cartier Reefs, Timor Sea	225	669	742
Kashiwa-Jima Island, Japan	224	768	738
Scott/Seringapatam Reefs, Western. Australia	220	593	725
Samoa Islands, Polynesia	211	852	694
Chesterfield Islands, Coral Sea	210	699	691
Sangalakki Island, Kalimantan, Indonesia	201	461	660
Bodgaya Islands, Sabah, Malaysia	197	516	647
Pulau Weh, Sumatra, Indonesia	196	533	644
Izu Islands, Japan	190	464	623
Christmas Island, Indian Ocean	185	560	606
Sipadan Island, Sabah, Malaysia	184	492	603
Rowley Shoals, Western Australia	176	505	576
Cocos-Keeling Atoll, Indian Ocean	167	528	545
North-West Cape, Western Australia	164	527	535
Tunku Abdul Rahman Is., Sabah	139	357	450
Lord Howe Island, Australia	139	395	450
Monte Bello Islands, W. Australia	119	447	382
Bintan Island, Indonesia	97	304	308
Kimberley Coast, Western Australia	89	367	281
Cassini Island, Western Australia	78	249	243
Johnston Island, Central Pacific	78	227	243
Midway Atoll, Pacific, U.S.A.	77	250	240
Rapa, Polynesia	77	209	240
Norfolk Island, Australia	72	220	223

Table 6. Coral Fish Diversity Index (CFDI) for regions or countries with figures for total reef and shore fish fauna (if known), and estimated fauna from CFDI regression formula.

Locality	CFDI	No. reef fishes	Estim. Reef fishes
Indonesia	504	2027	2019
Australia (tropical)	401	1627	1584
Philippines	387	?	1525
Papua New Guinea	362	1494	1419
S. Japanese Archipelago	348	1315	1359
Great Barrier Reef, Australia	343	1325	1338
Taiwan	319	1172	1237
Micronesia	315	1170	1220
New Caledonia	300	1097	1156
Sabah, Malaysia	274	840	1046
Northwest Shelf, Western Australia	273	932	1042
Mariana Islands	222	848	826
Marshall Islands	221	795	822
Ogasawara Islands, Japan	212	745	784
French Polynesia	205	730	754
Maldive Islands	219	894	813
Seychelles	188	765	682
Society Islands	160	560	563
Tuamotu Islands	144	389	496
Hawaiian Islands	121	435	398
Marquesas Islands	90	331	267

Endemism

Considering the broad dispersal capabilities via the pelagic larval stage of most reef fishes, it is not surprising that relatively few fish species are endemic to the Raja Ampat Islands. Six species are presently classified as endemics, but this status is provisional, pending further collecting in adjacent areas, particularly Halmahera, and the adjacent mainland of the Birdshead Peninsula. All of these species belong to families that exhibit parental care and presumably have brief larval stages. The "endemic" species are discussed in the following paragraphs.

Hemiscyllium freycineti (Quoy and Gaimard, 1824) (Hemiscyllidae) - A single specimen was photographed and collected at Kri Island (near Site 2b). The species was previously known on the basis of five specimens deposited at the Muséum National d'Histoire Naturelle, Paris. French naturalists collected these between 1817 and 1825 in the vicinity of Waigeo Island. The species is relatively common on shallow reefs of the Raja Ampat Islands, and is mainly seen at night.

Pseudochromis sp. (Pseudochromidae) – Several specimens were collected by the author during the present survey and also in 1998. The species is apparently new and closely related to *P. eichleri* Gill and Allen from the Philippines.

The species was commonly sighted on rubble bottoms at the base of steep slopes in about 18 to 20 m depth. It occurs alone or in pairs.

Apogon sp. 1 (Apogonidae) – About 15 individuals were sighted at Sites 12 and 37 at depths between 12–15 m. Three specimens were collected by spear. This new species will be named "leptofasciatus" and the description has been submitted for publication (Allen, in press b). It is most similar to *A. nigrocinctus* Smith and Radcliffe (northern Australia to the Philippines) and *A. jenkinsi* Evermann and Seale (Australia to Japan), both of which possess black markings on the dorsal fins and caudal-fin base that are also evident on the new species. However, adults of these species lack narrow stripes on the upper body, and *A. jenkinsi* also differs in having a black spot on each side of the nape. The new species further differs from these two species in having fewer developed rakers on the first branchial arch (18 versus 22–25).

Apogon sp. 2 (Apogonidae) – Only three specimens of this undescribed species were seen during the survey and a special effort was made to collect them. The fish were sighted in 45–50 m depth at Pef Island (Site 6). They were hovering a short distance above a *Halimeda*-covered rubble bottom among a large aggregation of *Apogon ocellicaudus*. This is another new species and will be named "oxygrammus" by the author (Allen, in press b). It differs from all known species in the genus on the basis of colour pattern (overall whitish with tapering black mid-lateral stripe that extends onto the caudal fin) and jaw dentition (enlarged teeth in relatively few rows).

Meiacanthus crinitus Smith-Vaniz, 1987 (Blenniidae) – This species was previously known on the basis of 11 specimens collected in 1979 from the vicinity of Batanta Island. During the present survey it was occasionally sighted throughout the survey area, usually on sheltered reefs with abundant live coral in 1–20 m depth. *Meiacanthus* possess poison fangs and are frequently mimicked by other fishes (Smith-Vaniz, 1976). Juveniles of the threadfin bream *Pentapodus trivittatus* (Nemipteridae) are very similar in appearance to *M. crinitus*, and Smith-Vaniz, Satapoomin, and Allen (in press) suggest that mimicry is involved.

Eviota sp. (Gobiidae) – This tiny, mid-water hovering goby was seen at seven sites (6, 19, 23, 35, 42–44), where it was locally common (16 specimens collected). It represents a new species and will be named "raja" (Allen, in press c). It is closely related to *E. bifasciata*, a sympatric species that is distributed across the Indo-Australian Archipelago. The two species differ in colour pattern, most notably the mid-lateral stripe (white in *E. bifasciata*, yellow in the new species) and the dark markings at the upper and lower caudal-fin base (horizontal streaks in *E. bifasciata*, vertically elongate spots in the new species). They also differ in counts for segmented rays in the second dorsal fin and lateral scale rows (usually 9 and 22 respectively for *E. bifasciata* and 10 and 25 in the new species).

Historical background

The seas surrounding the Raja Ampat Islands have long held a fascination for research scientists and explorers, although until recently there have been relatively few comprehensive observations or collections of coral reef fishes. Waigeo Island, in particular, was the focus of early French visits by several vessels including L'Uranie (1818-1819), La Coquille (1823), and L'Astrolabe (1826). Consequently, approximately 70 fish species were recorded, and Waigeo is an important type locality for a variety of fishes described mainly by Quoy and Gaimard (1824 and 1834), Lesson (1828–1830), and Cuvier and Valenciennes (1828–1849). Fishes that were originally described from Waigeo by early French researchers include such well-known species as the Black-tipped Shark (*Carcharhinus melanopterus*), Bluefin Trevally (*Caranx melampygus*), Bigeye Trevally (*Caranx sexfasciatus*), Semicircular Angelfish (*Pomacanthus semicirculatus*), and Sergeant Major (*Abudefduf vaigiensis*).

Following the early French explorations, most ichthyological activity was provided by Dutch researchers. The famous surgeon-naturalist Pieter Bleeker periodically received specimens from government agents and in 1868 published on a collection of Waigeo fishes that included 88 species. He added a further 12 species in subsequent papers. Albert Günther, the Curator of Fishes at the British

Museum, recorded 28 species from the island of Misool, during the cruise of the "Curacao" in 1865 (Günther, 1873). The Dutch ichthyologists Weber and de Beaufort were keenly interested in New Guinean freshwater and marine fishes and contributed to our knowledge of Raja Ampat fishes during the first half of the past century. The work of de Beaufort (1913), in particular, is the most extensive effort on Raja Ampat fishes until recent times, and includes accounts of 117 species based on 748 specimens. These were obtained by de Beaufort during a visit to the East Indies in 1909–1910, and were mainly collected at Waigeo in the vicinity of Saonek Island and Mayalibit Bay. Weber and De Beaufort and various coauthors including Koumans, Chapman, and Briggs included an additional 67 records from Waigeo and Misool in the *Fishes of the Indo-Australian Archipelago* (E.J. Brill, Leiden; 11 volumes published between 1921–1962). The Denison-Crockett South Pacific Expedition made small collections at Batanta and Salawati consisting of 29 species that were reported by Fowler (1939). The only other fish collection of note was that by Collette (1977) who reported 37 species from mangrove habitats on Misool and Batanta.

The present author made the first comprehensive observations of coral reef fishes during two brief visits during (1998–1999). Although the main focus was to document the freshwater fauna, approximately 20 hours of scuba and snorkel diving yielded observations of more than 500 coral reef fishes. These are incorporated in the list appearing as Appendix 4 of this report.

The known coral reef fauna until then consisted of approximately 236 species. This total includes the following 47 species that were not recorded either during

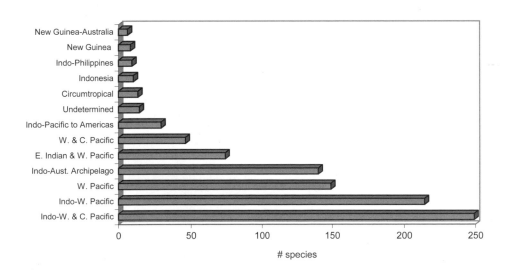

Figure 2. Zoogeographic analysis of Raja Ampat fishes

the 1998-1999 visits or during the current survey: *Moringua abbreviatus*, *M. javanicus*, *M. macrochir*, *Muraenichthys macropterus*, *M. gymnopterus*, *Enchelynassa canina*, *Ecidna delicatula*, *E. zebra*, *Gymnothorax boschi*, *G. meleagris*, *G. chilopilus*, *G. richardsoni*, *Ophichthys misolensis*, *Encheliophis homei*, *E. gracilis*, *Antennarius hispidus*, *A. nummifer*, *A. striatus*, *Hyporhamphus quoyi*, *Atherinomorus endrachtensis*, *Micrognathus brevirostris*, *Parascorpaena bandanensis*, *Scorpaenodes guamensis*, *Scorpaenopsis diabolis*, *Richardonichthys leucogaster*, *Centrogenys vaigiensis*, *Epinephelus undulosus*, *E. quoyanus*, *Apogon melas*, *Carangoides dinema*, *Lutjanus sebae*, *Gerres abbreviatus*, *Lethrinus nebulosus*, *Upeneus sulphureus*, *Halichoeres timorensis*, *Calotomus spinidens*, *Alticus saliens*, *Blenniella bilitonensis*, *Istiblennius edentulus*, *Paralticus amboinensis*, *Salarias guttatus*, *Synchiropus picturatus*, *Eviota zonura*, *Bathygobius fuscus*, *Gladigobius ensifer*, *Siganus vermiculatus*, and *Chelonodon patoca*. These species are either intimately associated with coral reefs or are fringe dwellers more commonly found in adjacent habitats (e.g. mangroves).

Overview of the Indonesian fish fauna

The Indonesian Archipelago is the world's premier area for marine biodiversity, mainly due to the extraordinary wealth of coral reef organisms. Until now, there have been scant details on the extent of the reef fish fauna, although there is nearly unanimous agreement that Indonesia has more coral fishes than any other region of the globe. Various family and generic revisions provide the best supporting evidence. For example, my own work on pomacentrids reveals about 145 species for the Indonesian Archipelago, easily more than any other region. This total represents about 40% of the world's total. In conjunction with Mohammad Adrim of Indonesia's National Oceanographic Institute (LON), I am now preparing a checklist of shallow (to 60 m depth) reef fishes of the Archipelago. Although species will continue to be added, the list currently contains 2,027 species. This total is compared with other leading countries in Table 9. Randall (1998) proposed the following factors to account for the extraordinary richness of the Indo-Australian region: 1) Sea temperatures have been very stable during

past glacial periods, preventing mass extinctions that occurred elsewhere in the Indo-Pacific; 2) The huge contiguous area of Indonesia and large number of island stepping-stones have formed a "buffer" against extinction; 3) The area is populated by numerous species with relatively short larval periods that are unable to cross deep-water oceanic barriers; 4) Some species have evidently evolved in peripheral regions and were subsequently transported to Indonesia via ocean currents, adding to the overall species richness; and 5) Lowered sea levels during past glacial periods have formed barriers that divided populations that eventually evolved into numerous geminate species pairs. Randall presented examples of 52 such pairings.

Judging from the present RAP and a number of additional surveys completed by the author since 1974, it appears that the area extending from central and northern Sulawesi to the western tip of Papua is possibly the richest area for reef fishes in Indonesia. The Raja Ampat Group are especially rich and seem to be a "cross-roads," containing faunal elements from Papua New Guinea and the Solomons to the east, Palau and the Philippines to the north, and the Moluccas and rest of the Indonesian Archipelago to the west.

Although most of Indonesia's reef fish fauna consists of widely distributed species (largely due to pelagic larval dispersal as already mentioned), there is a significant endemic element, consisting of at least 90 species. The endemics are scattered widely around the Archipelago, but there appear to be several "hotspots" including the Java Sea, Lesser Sunda Islands (especially the Komodo area), northern Sulawesi, and the Raja Ampat Islands (Allen and Adrim, in prep.). Most of the endemics, or about 83%, are included in just eight families; particularly prominent are the pseudochromids, blenniids, and pomacentrids. Well over half the species are confined to just nine genera. With the exception of the wrasse genus *Cirrhilabrus*, these are fishes that invariably exhibit parental egg care with a relatively short pelagic larval stage or in a few exceptional cases have completely abandoned the pelagic stage.

More than any other person, the great Dutch ichthyologist, Pieter Bleeker, is responsible for our present knowledge of Indonesian fishes. His extensive work in the country between 1842 and 1862 formed a sturdy foundation for following generations. The importance of his voluminous research can't be understated. Considering he was employed as an army surgeon during his 18-year stay in Indonesia, the extent of his ichthyological activity was remarkable. During an actual working career that spanned some 36 years Bleeker published 500 papers that include descriptions of an incredible number of new taxa: 529 genera and 1,925 species. Bleeker described approximately one of every six reef species presently recognized from Indonesia. His knowledge of the fauna, both fresh

Table. 9. The world's leading countries for reef fish diversity (from Allen, in press d).

Country	No. species
Indonesia	2,027
Australia	1,627
Philippines	1,525*
Papua New Guinea	1,494
Japan	1,315
Palau	1,254

* Estimated.

water and marine, was outstanding. Revisions of various groups of Indo-Pacific fishes by modern researchers frequently attest to Bleeker's uncanny intuition and astute understanding of natural relationships.

Discussion and Recommendations

There appears to be less impact from illegal fishing methods in the Raja Ampat Islands compared to other parts of Indonesia. The majority of sites visited were in good condition with an abundance of fishes of all sizes, although explosive damage was noted at seven locations. Villagers also informed us that cyanide is sometimes used to catch groupers and Napoleon wrasse for the live fish trade. Our observations of Napoleon Wrasse, a conspicuous indicator of fishing pressure, show that it is indeed heavily exploited, a typical situation in Indonesia. It was far more common at Milne Bay Province, Papua New Guinea, where illegal fishing methods are seldom used

(Table 10). With the exception of two large (> 100 cm) adults, most of the Napoleon Wrasse seen at the Raja Ampats were juveniles under 30 cm in length.

Table 11 presents the average number of species per site, number of sites where more than 200 species were observed, and the greatest number seen at a single site for all Marine RAP surveys to date. Despite a deliberate attempt to sample all habitats, including a relatively high proportion of sheltered environments where fishes are relatively poor, the Raja Ampats sites exhibited extraordinary faunal richness. A total of 200 or more species is generally considered by the author as the benchmark for an excellent fish count at a given site. This figure was obtained at 51% of Raja Ampat sites, well over twice as many times as its nearest Indonesian rival, the Togean-Banggai Islands.

Table 12 lists the 10 leading sites for fishes recorded by the author, during nearly 30 years of survey work in the Indo-Pacific region. Five of the ten all-time best sites were recorded during the present RAP.

Table 10. Frequency of Napoleon Wrasse (*Cheilinus undulatus*) sightings during five Marine RAPs.

Location	No. sites where seen	% of total sites	Approx. no. seen
Milne Bay, PNG – 2000	28	49.12	90
Milne Bay, PNG – 1997	28	52.83	85
Raja Ampat Islands – 2001	7	15.55	7
Togean/Banggai Islands – 1998	6	12.76	8
Weh Island, Sumatra – 1999	0	0.00	0
Calamianes Islands Philippines – 1998	3	7.89	5

Table 11. Comparison of site data for Marine RAP surveys 1997–2001.

Location	No. sites	Average spp./site	No. 200+ sites	Most spp. one site
Milne Bay, PNG	110	192	46 (42%)	270
Raja Ampat Islands	45	184	23 (51%)	283
Togean/Banggai Is., Sulawesi	47	173	9 (19%)	266
Calamianes Is., Philippines	21	158	4 (10.5%)	208
Weh Is., Sumatra	38	138	0	186

Table 12. G. Allen's 10 all-time best dive sites for fishes. MBP denotes Milne Bay Province while PNG denotes Papua New Guinea.

Rank	Location	No. spp.
1	Kri Island, Raja Ampat Is.	283
2	SE of Miosba Is., Fam Is., Raja Ampat Is.	281
3	Boirama Island, MBP, PNG	270
4	Irai Island, Conflict Group, MBP, PNG	268
5	Dondola Island, Togean Is., Indonesia	266
6	Keruo Island, Fam Is., Raja Ampat Is.	263
7	Pos II Reef, Menjangan Is., Bali, Indonesia	262
8	Equator Islands, Raja Ampat Is.	258
9	NW end Batanta Island, Raja Ampat Is.	246
10	Wahoo Reef, East Cape, MBP, PNG	245

Conservation

Every effort should be made to conserve the reefs of the Raja Ampat Islands. Although the present survey was by no means comprehensive, the very rich fauna that was documented over a relatively short period of time indicates an area of extraordinary fish diversity. The author has wide experience throughout the Indonesian Archipelago, and it is my opinion that no other area has as much potential for marine conservation. There are several reasons for this opinion:

- The exceptional habitat diversity and consequent rich fish fauna.

- Good condition of reefs compared to most other parts of Indonesia.

- A high aesthetic value based on the area's superb above-water and underwater scenery.

- A relatively low human population.

- Cultural values by indigenous Papuan people that are highly compatible with reef conservation.

- A rich and unique (many endemics) terrestrial fauna, which affords a rare opportunity to implement both marine and terrestrial conservation at the same time.

References

Allen, G. R. 1991. Damselfishes of the world. Aquarium Systems, Mentor, Ohio.

Allen, G. R. 1993. Reef fishes of New Guinea. Christensen Research Institute, Madang, Papua New Guinea Publ. No. 8.

Allen, G. R. 1997. Marine Fishes of tropical Australia and South-east Asia. Western Australian Museum, Perth.

Allen, G. R. 2001. Reef and Shore Fishes of the Calamianes Islands, Palawan Province, Philippines. *In*: Werner, T. B. and G. R. Allen , (eds.). A Rapid Marine Biodiversity Assessment of the Calamianes Islands, Palawan Province, Philippines. Bulletin of the Rapid Assessment Program 17, Conservation International, Washington, DC.

Allen, G. R. In press a. Reef Fishes of Milne Bay Province, Papua New Guinea. *In*: Allen, G. R. and T. B. Werner (eds.) A Rapid Marine Biodiversity Assessment of Milne Bay Province, Papua New Guinea. Bulletin of the Rapid Assessment Program 17, Conservation International, Washington, DC.

Allen, G. R. In press b. Two New Species of Cardinalfishes (Apogonidae) from the Raja Ampat Islands, Indonesia. Aqua, J. Ichthyol. Aquatic Biol.

Allen, G. R. In press c. Description of Two New Gobies (*Eviota*, Gobiidae) from Indonesian Seas. Aqua, J. Ichthyol. Aquatic Biol.

Allen, G. R. In press d. Indo-Pacific coral reef fishes as indicators of conservation hotspots. Proc. Ninth Intern. Coral Reef Symp.

Allen, G. R. and M. Adrim. In preparation. Reef fishes of Indonesia.

Bleeker, P. 1868. Notice sur la faune ichthyologique de l'ile de Waigiou. Versl. Akad. Amsterdam (2) II: 295–301.

Collette, B. B. 1977. Mangrove fishes of New Guinea. *In*: Teas, H. J. (ed.) Tasks for vegetation science. W. Junk Publishers, The Hague: 91–102.

Cuvier, G. and A. Valenciennes. 1828–1849. Histoire naturelle des poissons. 22 volumes. Paris.

de Beaufort, L. F. 1913. Fishes of the eastern part of the Indo-Australian Archipelago with remarks on its zoogeography. Bijd. Neder. Dierk., Amsterdam 19: 95–163.

Fowler, H. W. 1939. Zoological results of the Denison-Crockett South Pacific Expedition for the Academy of Natural Sciences of Philadelphia, 1937–1938. Part III. – Fishes. Proc. Acad. Nat. Sci. Philadelphia 91: 77–96.

Kuiter, R. H. 1992. Tropical reef fishes of the Western Pacific-Indonesia and adjacent waters. Percetakan PT Gramedia Pustaka Utama, Jakarta.

Lesson, R. P. 1828. Description du noveau genre Ichthyophis et de plusierus espéces inédites ou peu connues de poissons, recueillis dans le voyage autour du monde de la Corvette "La Coquille." Mem. Soc. Nat. Hist. Paris v. 4: 397–412.

Lesson, R. P. 1830–31. Poissons. *In*: Duperrey, L. (ed.) Voyage austour du monde, …, sur la corvette de La Majesté La Coquille, pendant les années 1822, 1823, 124 et 1825…, Zoologie. Zool. v. 2 (part 1): 66–238.

Myers, R. F. 1989. Micronesian reef fishes. Coral Graphics, Guam.

Quoy, J. R. C. and J. P. Gaimard. 1824. Voyage autour du monde, Enterpris par ordre du Roi exécuté sur les corvettes de S. M. "L'Uranie" et "La Physicienne" pendant les années1817, 1818, 1819, et 1820, par M. Louis de Freycinet. Zool. Poissons: 183–401.

Quoy, J. R. C. and J. P. Gaimard. 1834. Voyage de découvertes de "L'Astrolabe" exécuté par ordre du Roi, pendant les années1826–1829, sous le commandement de M. J. Dumont d'Urville. Poissons III: 647–720.

Randall, J. E., G. R. Allen, and R. C. Steene. 1990. Fishes of the Great Barrier Reefand Coral Sea. Crawford House Press, Bathurst (Australia).

Randall, J. E. and Randall, H. A., 1987. Annotated checklist of fishes of Enewetak Atoll and Other Marshall Islands. *In*: Vol 2. The natural History of Enewetak Atoll. Office of Scientific and Technological Information U.S. Dept. of Energy: 289–324.

Randall, J. E. 1998. Zoogeography of shore fishes of the Indo-Pacific region. Zool Stud. 37: 227–268.

Smith-Vaniz, W. F. 1976. The saber-toothed blennies, tribe Nemophini (Pisces: Blenniidae). Monograph 19, Acad. Nat. Sci. Philadelphia.

Springer, V. G. 1982. Pacific plate biogeography with special reference to shorefishes. Smith Contrib Zool. 367: 1–182.

Winterbottom, R., A. R. Emery, and E. Holm. 1989. An annotated checklist of the fishes of the Chagos Archipelago, Central Indian Ocean. Roy. Ontario Mus. Life. Sci. Contrib. 145:1–226.

Chapter 4

A Basic Stock Assessment of Economically Important Coral Reef Fishes of the Raja Ampat Islands, Papua Province, Indonesia

La Tanda

Ringkasan

- Pendugaan stok ikan-ikan karang dilakukan di Kepulauan Raja Ampat, Kabupaten Sorong, Papua pada bulan Maret – April 2001

- Sebanyak 196 spesies, mewakili 59 genus dan 19 famili dikategorikan sebagai kelompok ikan karang target untuk perikanan. Dua spesies ikan yang banyak dijumpai adalah *Pterocaesio pisang* dan *Caesio cuning*.

- Penghitungan spesies ikan target pada tiap lokasi berkisar antara 14–72 (rata-rata = 42.01±1,78). Jumlah individu ikan target berkisar 79–2760 (rata-rata = 810,64±94,18). Perkiraan biomassa ikan target di satu lokasi berkisar antara 27,09–1031,8 ton/km² (rata-rata = 208,97±27,83 ton/km²).

- Rata-rata total pendugaan biomassa ikan-ikan target di semua lokasi survei di Kep. Raja Ampat sangat tinggi dibandingkan kawasan lain di "Coral Triangle" yang pernah disurvei, seperti Propinsi Milne Bay (Papua New Guinea), Kepulauan Togean-Banggai (Indonesia) dan Kepulauan Calamianes (Filipina).

Summary

- A stock assessment of coral reef fishes was undertaken at the Raja Ampat Islands, Sorong District, Papua Province in March-April 2001.

- A total of 196 species, representing 59 genera and 19 families, were classified as target species for reef fisheries. Two species, *Pterocaesio pisang* and *Caesio cuning* (Family Caesionidae) were particularly abundant.

- Counts of target species at individual sites ranged between 14–72 (mean = 42.0 ± 1.8). Counts of individual target fishes at a single site ranged between 79 to 2760 (mean = 810.6 ± 94.18). Estimated target fish biomass at a single site ranged between 27.09–1031.8 ton/km² (mean =208.97 ± 27.83 ton/km²).

- The mean total biomass estimate for sites in the Raja Ampat Islands is considerably greater than for other previously sampled areas in the "coral triangle" including Milne Bay Province (Papua New Guinea), Togean-Banggai Islands (Indonesia), and Calamianes Islands (Philippines).

Introduction

Coral reefs are vitally important to Indonesian communities. More than 30 million people live in close proximity to the sea and rely heavily on marine resources for food and income. Reefs provide necessary shelter and breeding grounds for a variety of marine organisms. They also form a protective barrier, sheltering coastal villages from the onslaught of oceanic waves. Moreover, coral organisms represent a valuable source of medicinal ingredients, and are also used in the manufacture of ornamental jewellery. Coral reefs are often an important source of extra income through benefits derived from eco-tourism. Unfortunately, Indonesian coral reefs are currently being over-exploited at an alarming rate. Destructive fishing methods, particularly the use of explosives and cyanide, are especially prevalent. Previous survey results revealed that only about 6% percent of Indonesian reefs were classified as being in good condition, and about 40% were classified in poor condition (Suharsono, 1999). The simple truth is that reef habitat is rapidly disappearing. As Indonesia's population continues to spiral, there is increased pressure on marine resources, especially the species-rich and highly productive coral reef environment. Therefore, it is extremely important to identify and protect the nation's remaining areas of rich coral reef biodiversity. Hopefully, these will form a network that will ensure the long-term survival of Indonesia's extraordinary wealth of reef organisms.

The primary goal of the current study was to collect essential information on marine resources that are utilized by local communities in the Raja Ampat Islands. Hopefully this data will assist in the management of sustainable coral reef resources so that they can be wisely exploited by future generations.

Material and Methods

Observations were carried out in March-April 2001 at 45 sites in the Raja Ampat Islands. Data were collected visually while SCUBA diving and recorded with pencil on waterproof plastic paper. The visual census methodology outlined by Dartnall and Jones (1986) was employed with some modifications. Observations were made while slowly swimming along a 10-m wide "corridor" centered on a 100-m tape measure that was placed on the bottom along a predetermined depth contour, forming a survey area of approximately 1000 m² per transect. The time spent on each transect ranged from 20–35 minutes. Data were collected for three transects at most sites: deep (18–20 m), moderate (10–13 m), and shallow (4–6 m).

Target species are defined as edible fishes that live on or near coral reefs. They include fishes that are

important in both commercial and artisenal catches. Numbers of individuals and average size were recorded for every target species that was observed. Data for numbers of individuals were obtained by actual count, except when fish occurred in large schools, in which case rough estimates were made to the nearest 50–100 fish. Average sizes were estimated to the nearest five cm.

These data were used to calculate fish biomass (expressed in ton/km²). Average length was converted into weight using the cubic law:

Weight = 0.05(Length)³   **Units: Weight in grams (g)**
                              Length in centimeters (cm)

Identification references included Kuiter (1993), Allen (1997), and Matsuda and Allen (1987).

Results and Discussion

A total of 196 species representing 56 genera and 19 families were recorded (Appendix 5). Slightly over half of this total is composed of the families Acanthuridae, Serranidae, Scaridae, Lutjanidae, and Caesionidae (Fig. 1).

The most commonly observed target species (percentages of occurrence in parentheses) were as follows: *Parupeneus multifasciatus* (90.91), *Ctenochaetus striatus* and *Chlorurus bleekeri* (84.09), *Cheilinus fasciatus* (81.82), *Parupeneus barberinus* (79.55), *Zebrasoma scopas* (68.18), *Acanthurus pyroferus*, *Scolopsis bilineatus* and *Siganus corallinus* (65.91), *Monotaxis granduculus* and *Plectorhinchus polytaenia* (63.64), *Pterocaesio pisang* and *Naso hexacanthus* (56.82).

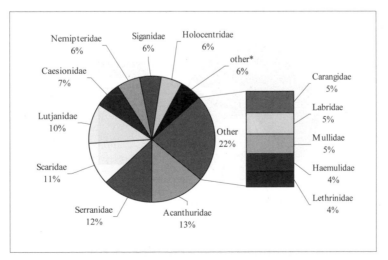

Figure 1. Percentages of target species belonging to different families. (*Other families are those with less than five species recorded.)

Caesionids, commonly known as fusiliers, were the most abundant fishes at most sites. Indeed, they comprised nearly 66% of the total 'target' fish count (Fig. 2). Given their dominance regarding number of individuals, it was not surprising that caesionids represented the largest segment of the total biomass as well (Fig. 3). Other important families in this regard included Acanthuridae, Lutjanidae, and Scaridae (Fig. 3). Families with less than five species seen during the entire survey (labeled as "other" in Figs. 1, 2, and 3) contributed only 2% of the overall biomass.

Summary of data for Raja Ampat sites (refer to Table 1): Counts of target *species* ranged from 5–37 (mean = 20.77 ± 1.76) for deep transects, from 9–39 (mean = 25.10 ± 1.19) for moderate transects and 14–38 (mean = 25.26 ± 0.99) for shallow transects. Therefore, the data indicate that target species are slightly more numerous in shallow water. If the counts for target species from the various depth zones at each site are combined, the numbers range from 14–72 (mean = 42.02 ± 1.78).

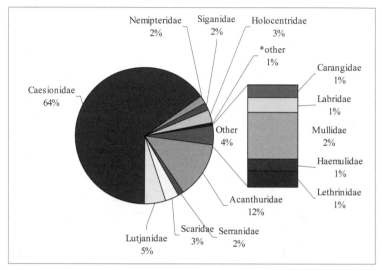

Figure 2. Composition of fish counts for Raja Ampat families. (*Other families are those which had less than five species recorded.)

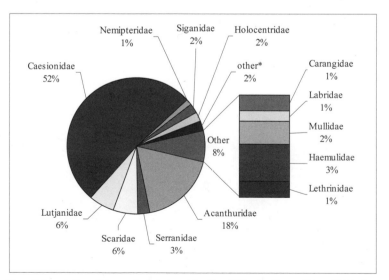

Figure 3. Composition of biomass for Raja Ampat families. (*Other families are those which had less than five species recorded.)

Table 1. Summary of coral reef fish stock assessment.*

Site	Transect 1 (18-20 m)			Transect 2 (10-13 m)			Transect 3 (4-6 m)			Site Total			Most abundance species
	No. target species	Approx. fish count	Biomass (ton/km2)	No. target species	Approx. fish count	Biomass (ton/km2)	No. target species	Approx. fish count	Biomass (ton/km2)	No. target species	Approx. fish count	Biomass (ton/km2)	(% of total fish count)
27	12	130	84.63	15	84	67.25	18	103	94.31	27	317	82.06	*Caesio cuning* (61.51 %)
28	---	---	---	22	308	324.52	31	378	208.63	39	686	266.58	*Pterocaesio tile* (40.09 %)
29	---	---	---	17	307	44.87	28	602	280.07	35	909	162.47	*Lutjanus biguttatus* (55.01 %)
30	18	122	80.67	23	226	162.42	---	---	---	32	348	121.55	*Acanthurus blochii* (21.84 %)
31	26	151	66.92	19	89	104.39	22	655	225.83	42	893	132.26	*Pterocaesio pisang* (64.38 %)
32	---	---	---	29	1162	622.94	19	310	107.44	39	1472	365.19	*Pterocaesio pisang* (33.96 %)
33	11	60	35.82	29	553	375.27	21	113	87.36	43	726	166.15	*Pterocaesio marri* (34.44 %)
34	---	---	---	24	315	225.79	26	297	240.99	35	612	233.39	*Pterocaesio marri* (44.93 %)
35	5	6	3.76	23	66	48.32	14	37	29.17	30	109	27.09	*Caesio cuning* (22.94 %)
36	28	387	413.96	9	125	88.75	---	---	---	31	512	251.35	*Caesio cuning* (34.18 %)
37	---	---	---	31	158	109.61	29	500	89.41	44	658	99.51	*Caesio lunaris* (45.89 %)
38	18	82	39.89	15	113	44.61	21	111	55.91	37	306	46.8	*Caesio cuning* (42.48 %)
39	36	223	235.61	25	185	140.22	35	477	129.94	63	884	168.14	*Caesio teres* (39.59 %)
40	---	---	---	21	92	2305	32	281	73.84	44	373	48.44	*Caesio caerulaurea* (40.21 %)
41	14	48	28.05	21	68	29.69	26	756	487.92	43	872	181.89	*Caesio cuning* (42.43 %)
42	27	501	160.62	31	139	141.15	25	167	111.66	45	640	100.59	*Pterocaesio pisang* (31.21)
43	---	---	---	38	617	334.69	28	1198	487.67	46	1822	412.58	*Pterocaesio pisang* (41.16 %)
44	32	332	160.41	25	482	178.19	32	176	177.82	62	980	170.8	*Pterocaesio pisang* (35.71 %)
N	26	26	26	41	41	41	39	39	39	44	44	44	Ranking of most abundant species
Min	5	6	3.76	9	20	23.05	14	37	29.17	14	79	27.09	
Max.	37	1216	800.66	39	1544	1256.29	38	1198	690.21	72	2760	1031.8	1. *Caesio cuning* (34 sites)
Avg.	20.77	304.42	180.39	25.10	389.17	249.56	25.26	326.85	175.31	42.02	810.64	208.97	2. *Pterocaesio pisang* (26 sites)
S.E.	1.76	65.00	37.25	1.19	55.73	39.04	0.99	39.85	25.77	1.78	94.18	27.83	

*continued on page 62.

| | Transect 1 (18-20 m) | | | Transect 2 (10-13 m) | | | Transect 3 (4-6 m) | | | Site Total | | | Most abundance species |
Site	No. target species	Approx. fish count	Biomass (ton/km2)	No. target species	Approx. fish count	Biomass (ton/km2)	No. target species	Approx. fish count	Biomass (ton/km2)	No. target species	Approx. fish count	Biomass (ton/km2)	(% of total fish count)
27	12	130	84.63	15	84	67.25	18	103	94.31	27	317	82.06	Caesio cuning (61.51 %)
28	----	----	----	22	308	324.52	31	378	208.63	39	686	266.58	Pterocaesio tile (40.09 %)
29	----	----	----	17	307	44.87	28	602	280.07	35	909	162.47	Lutjanus biguttatus (55.01 %)
30	18	122	80.67	23	226	162.42	----	----	----	32	348	121.55	Acanthurus blochii (21.84 %)
31	26	151	66.92	19	89	104.39	22	655	225.83	42	893	132.26	Pterocaesio pisang (64.38 %)
32	----	----	----	29	1162	622.94	19	310	107.44	39	1472	365.19	Pterocaesio pisang (33.96 %)
33	11	60	35.82	29	553	375.27	21	113	87.36	43	726	166.15	Pterocaesio marri (34.44 %)
34	----	----	-	24	315	225.79	26	297	240.99	35	612	233.39	Pterocaesio marri (44.93 %)
35	5	6	3.76	23	66	48.32	14	37	29.17	30	109	27.09	Caesio cuning (22.94 %)
36	28	387	413.96	9	125	88.75	----	----	----	31	512	251.35	Caesio cuning (34.18 %)
37	----	----	----	31	158	109.61	29	500	89.41	44	658	99.51	Caesio lunaris (45.89 %)
38	18	82	39.89	15	113	44.61	21	111	55.91	37	306	46.8	Caesio cuning (42.48 %)
39	36	223	235.61	25	185	140.22	35	477	129.94	63	884	168.14	Caesio teres (39.59 %)
40	----	----	----	21	92	2305	32	281	73.84	44	373	48.44	Caesio caerulaurea (40.21 %)
41	14	48	23.05	21	68	29.69	26	756	487.92	43	872	181.89	Caesio cuning (42.43 %)
42	27	501	160.62	31	139	141.15	25	167	111.66	45	640	100.59	Pterocaesio pisang (31.21)
43	----	----	----	38	617	334.69	28	1198	487.67	46	1822	412.58	Pterocaesio pisang (41.16 %)
44	32	332	160.41	25	482	178.19	32	176	177.82	62	980	170.8	Pterocaesio pisang (35.71 %)
N	26	26	26	41	41	41	39	39	39	44	44	44	Ranking of most abundant species
Min	5	6	3.76	9	20	23.05	14	37	29.17	14	79	27.09	
Max.	37	1216	800.66	39	1544	1256.29	38	1198	690.21	72	2760	1031.8	1. Caesio cuning (34 sites)
Avg.	20.77	304.42	180.39	25.10	389.17	249.56	25.26	326.85	175.31	42.02	810.64	208.97	2. Pterocaesio pisang (26 sites)
S.E.	1.76	65.00	37.25	1.19	55.73	39.04	0.99	39.85	25.77	1.78	94.18	27.83	

Numbers of individuals for various target fishes ranged from 6 to 1,216 (mean = 304 ± 65.0) on deep transects, from 20 to1,544 (mean = 389.2 ± 55.73) on moderate transects and from 37 to 1,198 (mean = 326.8 ± 39.85) on shallow transects. Thus, the data show there is a greater abundance of target fish on moderate-depth transects.

The estimated target fish biomass ranged from 3.76 to 800.66 ton/km^2 (mean = 180.4 ± 37.25ton/km^2) on deep transects, from 23.05 to 1256.29 ton/km^2 (mean = 249.6 ± 39.04 ton/km^2) on moderate transects and 29.17 to 690.21 ton/km^2 (mean = 175.3 ± 25.77 ton/km^2) on shallow transects. Total biomass for each site ranged from 27.1 to 1,031.8 ton/km^2 (mean = 209 ± 27.8 ton/km^2).

The best-ranked sites for the three data categories are indicated in Table 2. Biomass is the most important category from a fisheries perspective. The small group of islets (Melissa's Garden, site 18) off the southeastern side of North Fam Island was by far the richest area for target species biomass. The main fishes responsible for its extraordinary value were fusiliers (*Caesio cuning*, *Pterocaesio tesselata*, and *Pterocaesio pisang*) and surgeonfishes (*Acanthurus mata*, *Naso hexacanthus*, and *Naso vlamingi*). These species formed large shoals over a predominately live-coral substratum at depths between about 10–20 m.

Figure 4 compares the Raja Ampats with three other areas previously surveyed by Conservation International. The mean estimated biomass for all sites was significantly greater than that recorded at Milne Bay (Papua New Guinea), the Togean-Banggai Islands (Indonesia), and the Calamianes Islands (Philippines). It is particularly notable that the figure for the Raja Ampat Islands is three times greater than that of the Togean-Banggai Islands. The order of magnitude of this difference no doubt reflects the much lower fishing pressure in the Raja Ampats compared to other parts of Indonesia.

The Serranidae was the most speciose of the target families in the Raja Ampat Islands. A total of 52 species were recorded, including 25 that are considered as target species. The larger members of this family are among the most sought after of fishes by Indonesian fishers due to their size and good-eating qualities. Consequently, they have a very high economic value.

Table 2. Best sites for target fishes in three data categories at the Raja Ampat Islands.

Rank	Number of target species		Approximate fish count		Estimated biomass (ton/km2)	
	Site	Value	Site	Value	Site	Value
1	14	72	18	2760	18	1031.80
2	39	63	14	2237	14	607.91
3	44	62	21	2101	21	530.88
4	16	61	16	1989	17	429.16
5	2a	54	43	1822	43	412.58
6	25	53	10	1788	16	386.98
7	24	52	32	1472	32	365.19
8	17	50	7	1308	7	324.26

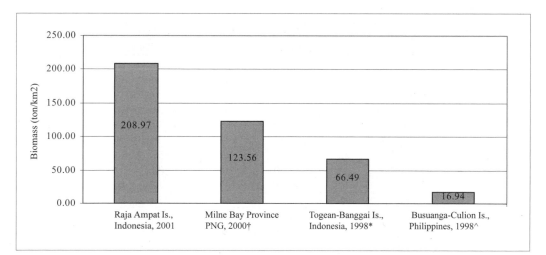

Figure 4. Comparison of mean "site total" biomass for past and present CI RAP surveys. (sources: †Allen et al., 2000; *La Tanda, 1998; ^Ingles, 1998)

Sites in Raja Ampat Islands appear to have a significantly greater mean density of target serranids (Fig. 5). However, typical of other locations in the Coral Triangle, the average size of groupers in the Raja Ampats was relatively small (about 25 cm). However, fishes (mainly *Plectropomus* and *Variola*) of 40 cm or longer were encountered at many sites. The average value at the Calamianes Islands, Philippines, was typically less than 20 cm (Ingles, 1998) and slightly greater than 20 cm at sites at the Togean-Banggai Islands, Indonesia (La Tanda, 1998). By contrast, the mean value for Milne Bay Province, Papua New Guinea, exceeded 30 cm

(Allen et al., in press). The density of Raja Ampat groupers per 1000 m² ranged between 0.67 to 17.0 individuals with the highest density recorded at sites 23 and 44 (Table 3).

The Napoleon Wrasse was rarely observed during the present survey. Allen (chapter in this report) noted its presence at only seven sites, and in most cases individuals were under 30 cm in length. This species, which may reach a size well in excess of 1 m, is intensively harvested for export in the live-fish restaurant trade. It is sold locally to Sorong merchants for about $20 USD/kg.

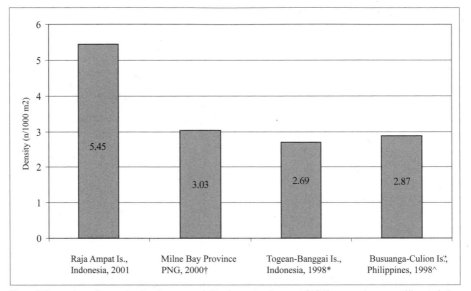

Figure 5. Comparison of mean density of groupers at sites for past and present CI RAP surveys. (sources: †Allen et al., in press; * La Tanda, 1998; ^ Ingles, 1998).

Table 3. Density of groupers for 44 sites at the Raja Ampat Islands.

Site No.	Density (ind/1000m)²	Site No.	Density (ind/1000m)²	Site No.	Density (ind/1000m)²
1	11.3	16	5.00	31	1.67
2a	8.00	17	5.50	32	2.50
2b	0.67	18	4.00	33	5.67
3	4.00	19	1.33	34	3.50
4	10.00	20	16.50	35	1.67
6	9.00	21	2.50	36	2.50
7	6.00	22	4.50	37	5.00
8	2.00	23	17.00	38	3.67
9	0.00	24	4.00	39	1.33
10	3.67	25	5.00	40	1.50
11	2.00	26	2.33	41	4.33
12	5.33	27	2.33	42	4.00
13	2.33	28	7.00	43	13.50
14	11.00	29	5.50	44	17.00
15	7.67	30	6.00		

Conclusions

- Target species diversity observed during this survey was greatly variable according to site, but was generally higher than observed by the author at other reef areas in Indonesia. Local coral reef diversity was augmented by adjacent ecosystems, particularly mangroves and seagrass beds.

- The present RAP survey and work that is planned for the future by CI fill a critical gap of knowledge and represent important steps in the overall conservation of the area's marine biodiversity. These studies are not designed to include just biological parameters, but also strongly emphasize the exploitation of marine resources and the resulting socio-economic implications. The overall goal of these activities is to create a workable balance between conservation and the utilization of resources.

References

Ad rim, M. and Yahmantoro. 1993. Studi pendahuluan terhadap fauna ikan karang di perairan P.P. Tiga, Sulawesi Utara. Wisata Bahari Pulau-Pulau Tiga (Tundonia, Tengah , Paniki) Sulawesi Utara. Lembaga Ilmu Pengetahuan Indonesia, Puslitbang Oseanologi. Proyek Penelitian dan Pengembangan Sumberdaya Laut. Jakarta: 29–44.

Allen, G.R. 1997. Marine Fishes of Tropical Australia and South-East Asia. Western Australian Museum, Perth. 292 pp.

Allen, G.R. and R.C. Steene. Indo-Pacific Coral Reef Field Guide. Tropical Reef Research, Singapore. 378 pp.

Allen, M., J. Kinch and T.B. Werner. In press. A Basic Stock Assessment of the Coral Reef Resources of Milne Bay Province, Papua New Guinea, Including a Study of Utilization at the Artisan Level. *In*: G.R. Allen and P. Seeto (eds.). A Rapid Marine Biodiversity Assessment of Milne Bay Province, Papua New Guinea second survey. Bulletin of the Rapid Assessment Program, Conservation International, Washington, DC.

Dartnall, H.J. and M. Jones. 1986. A manual of survey methods of living resources in coastal areas. ASEAN-Australia Cooperative Programme on Marine Science Handbook. Australian Institute of Marine Science, Townsville. 167 pp.

Kuiter, R.H. 1992. Tropical Reef-Fishes of the Western Pacific. Indonesia and Adjacent Waters. Gramedia, Jakarta. 314 pp.

La Tanda. 1998. Species composition, distribution and abundance of coral fishes in the Togean and Banggai Islands. *In*: Werner, T.B., G.R. Allen , and S.A. McKenna (eds.). A Rapid Marine Biodiversity Assessment of the Togean and Banggai Islands, Sulawesi, Indonesia. Bulletin of the Rapid Assessment Program, Conservation International, Washington, DC.

Suharsono, R. Sukarno, M. Adrim, D. Arief, A. Budiayanto, Giyanto, A. Ibrahim and Yahmantoro, 1995. Wisata Bahari Kepulauan Banggai. Lembaga Ilmu Pengetahuan Indonesia, Puslitbang Oseanologi. Proyek Penelitian dan Pengembangan Sumberdaya Laut. Jakarta. 44 pp.

Werner, T.B. and G.R. Allen (eds.). 1998. A Rapid Marine Biodiversity Assessment of Milne Bay Province, Papua New Guinea. Bulletin of the Rapid Assessment Program 17, Conservation International, Washington, DC.

Chapter 5

Condition of Coral Reefs at the Raja Ampat Islands, Papua Province, Indonesia

Sheila A. McKenna, Paulus Boli, and Gerald R. Allen

Ringkasan

- Kondisi terumbu karang adalah istilah mengenai kondisi "kesehatan" umum suatu lokasi berdasarkan penilaian beberapa parameter kunci, antara lain kerusakan lingkungan yang disebabkan manusia dan alam, dan keanekaragaman hayati umum ditentukan berdasarkan kelompok indikator utama (karang dan ikan).

- Terdapat 45 lokasi di Raja Ampat yang kondisi terumbu karangnya diteliti, termasuk di sekitar Pulau-pulau Batanta, Kri, Fam, Gam, Wayag, dan Kelompok Batang Pele.

- Nilai Indeks Kondisi Karang atau Reef Condition Index (RCI) dihitung untuk tiap lokasi. Nilai itu berasal dari tiga komponen yang diukur, yaitu keragaman karang, keragaman ikan, dan kerusakan relative akibat kegiatan manusia dan penyebab alami. Katagori yang terakhir juga memperhitungkan persentase tutupan karang hidup.

- Nilai hipotesa maksimum RCI untuk karang yang masih asli adalah 300; nilai RCI beguna untuk memahami kondisi karang sehingga dapat dibandingkan antar lokasi. Berdasarkan nilai RCI-nya, lokasi -lokasi tersebut dapat dikelompokkan sebagai luar biasa, sangat bagus, bagus, sedang, jelek, dan sangat jelek. Frekuensi di lokasi-lokasi Raja Ampat adalah sebagai berikut: Luar Biasa (0), Sangat bagus (17), bagus (10), sedang (10), buruk (6), dan sangat buruk (2).

- Nilai RCI tertinggi (242,90) terdapat di pulau Wofah, sebelah barat daya Waigeo. Daerah geografi utama yang memiliki nilai rata-rata RCI tertinggi adalah Kepulauan Fam (212,48), Batang Pele hingga Pulau Yeben (210,46), dan pulau Batanta Wai (207,93).

- Rata-rata nilai RCI untuk Raja Ampat (196,54 $\pm$ 4,89) sedikit lebih rendah walau hampir menyamai nilai di Milne Bay (199,32 $\pm$ 3,76). Nilai tersebut secara signifikan lebih tinggi dibandingkan nilai di Kepulauan Togean-Banggai, Indonesia (179,87 $\pm$ 4,02), satu-satunya daerah lain yang data RCI-nya tersedia.

- Ancaman atau kerusakan pada karang yang paling sering teramati di seluruh kawasan survei adalah tekanan penangkapan ikan, terdapat pada 73% dari seluruh lokasi. Tingkat tekanan pemanfaatan ikan tergolong kecil pada lokasi-lokasi tersebut kecuali lokasi 11. Bukti pengeboman ikan ditemukan pada 13,3% dari seluruh lokasi.

- Tekanan terbesar kedua yang teramati pada karang adalah pengendapan. Ditemukan pada 35,6% dari seluruh lokasi survei.

- Eutrofikasi/polusi teramati pada 17,8% dari seluruh lokasi survei. Lokasi-lokasi tersebut letaknya dekat dengan pantai.

- Sangat sedikit pemangsa karang teramati di daerah survey. Bintang laut (*Acanthaster planci*) atau bukti kehadirannya terlihat sebanyak 6,7% dari lokasi survei. Moluska pemakan karang *Drupella cornus* hanya terlihat pada satu lokasi terumbu.

- Sedikit sekali ditemukan (5,6% dari seluruh lokasi survey) pemutihan karang di lokasi terumbu. Tidak ada kerusakan serius atau kondisi pemutihan masal pada semua terumbu yang disurvei. Menariknya, menurut masyarakat lokal tidak pernah ada pemutihan masal di daerah penelitian dimasa lalu.

- Hanya ditemukan sedikit (4,5% dari lokasi survei) penyakit karang pada beberapa terumbu tapi tidak tersebar luas. Penyakit karang yang umum ditemukan adalah penyakit pita hitam dan putih.

- Satwa laut karismatik yang teramati selama survey adalah ikan pari, hiu, penyu, dan paus.

Summary

- Reef condition is a term pertaining to the general "health" of a particular site as determined by assessment of key variables including natural and human-induced environmental damage, and general biodiversity as defined by major indicator groups (corals and fishes).

- Reef condition was assessed at 45 sites in the Raja Ampat Islands, including reefs off of the islands of Batanta, Kri, Fam, Gam, Wayag, and Batang Pele Group.

- A Reef Condition Index (RCI) value was calculated for each site. The value is derived from three equally weighted components: coral diversity, fish diversity, and relative damage from human and natural causes. The latter category also incorporated the percentage of live coral cover.

- The hypothetical maximum RCI for a pristine reef is 300; RCI values are useful for interpreting reef condition and comparing sites. Depending on their RCI, sites can be classified as extraordinary, excellent, good, moderate, poor, and very poor. The frequency of Raja Ampat sites was as follows: extraordinary (0), excellent (17), good (10), moderate (10), poor (6), and very poor (2).

- The highest RCI value (242.90) was recorded for Wofah Island off of southwest Waigeo. Major geographic areas with the highest mean RCIs include Fam Islands (212.48), Batange Pele to Pulau Yeben (210.46), and Batanta Wai Island (207.93).

- The mean RCI value for Raja Ampat (198.04 ± 4.89) is slightly less although similar to that of Milne Bay (199.32 ± 3.76). These values are significantly greater than the value (179.87 ± 4.02) obtained for the Togean-Banggai Islands of Indonesia, the only other area for which RCI data is available.

- The most frequent threat or damage observed on reefs throughout the survey region was fishing pressure, recorded at 73% of the sites. The extent of fishing pressure was slight at these sites with the exception of site 11. Evidence of blast fishing was found at 13.3% of the sites.

- The second most frequent stressor on the reefs was siltation. This was evident at 37.8% of the sites surveyed.

- Eutrophication/pollution was observed at 17.8% of the sites surveyed. These sites were all located close to shore.

- Very few coral predators were observed throughout the sites surveyed. Crown-of-thorns starfish (*Acanthaster planci*) or evidence of its presence was seen at 6.7% of the sites surveyed. The coral-feeding mollusc *Drupella cornus* was only seen at one reef site.

- Very low incidence (5.6% of the sites surveyed) of coral bleaching was recorded on reef sites. No serious or mass bleaching events noticed at any of the reef sites surveyed. Interestingly, locals recall no past major bleaching events on the reef areas studied.

- Minor incidence (4.5% of sites surveyed) of coral pathogens on several reef sights with no major coral disease outbreaks was noted. Common coral diseases noted included white and black band disease.

- Charismatic marine fauna observed during the survey were manta rays, sharks, turtles, and short finned pilot whales.

Introduction

Indonesia holds approximately 50,000 km² of coral reef habitat stretching 5,000 km from east to west (Spalding et al. 2001). The reefs are identified as one of the most biologically diverse and threatened (Bryant et al. 1998). Interestingly, some reefs of the vast Indonesian archipelago remain undescribed. One such region lies off the coast of Papua. This study focused on the reefs off of Bird Head's peninsula, mainly the Raja Ampat Islands. Forty-five reef sites and their condition are described for the major geographica areas surveyed. These areas included Fam Islands, Batang Pele to Pulau Yeben, Batanta – Wai Island, Kawe – Wayag Islands, Alyui Bay, Gam-Mansuar, and Mayalibit Bay (see map).

Stressors to coral reefs are often multiple and synergistic. Regardless of the cause, damage or stress to a reef can lead to loss of habitat ecosystem functioning and ultimately to the tragic extinction of species. Reef condition is a term used to indicate the general "health" of a particular site as determined by an assessment of several variables. These include biodiversity as defined by key indicator groups (corals and fishes) and relative damage or stress from human and natural causes. The latter category also incorporates the percentage of live coral cover.

Materials and Methods

Reef Condition Index

RAP surveys provide an excellent vehicle for rapid documentation of biodiversity of previously unstudied sites. They also afford an opportunity to issue a "report card" on the status or general condition of each reef site. However, this task is problematical. The main challenge is to devise a rating system that is not overly complex, and accurately reflects the true situation, thus providing a useful tool for comparing all sites for a particular RAP or for comparing sites in different regions. CI's Reef Condition Index (RCI) has evolved by trial and error, and although not yet perfected, shows promise of meeting these goals. The present method was trialed during the Togean-Banggai RAP in 1998 and the second survey of Milne Bay in 2000. Data are now available for 104 sites, including those from the current survey. Basically, the data consist of three equal components: fish diversity, coral diversity, and condition factors.

Fish diversity component – Total species observed at each site. A hypothetical maximum value of 280 species is utilized to achieve equal weighting. Therefore, the species total from each site is adjusted for equal weighting by multiplying the number of species by 100 and dividing the result by 280.

Coral diversity component – Total species observed at each site. A hypothetical maximum value of 130 species is utilized to achieve equal weighting. The species total from each site is adjusted for equal weighting by multiplying the number of species by 100 and dividing the result by 130.

Reef condition component - This is the most complex part of RCI formula and it is therefore instructive to give an example of the data taken from an actual site (site 1):

Parameter	1	2	3	4
1. Explosive/Cyanide damage				X
2. Net damage				X
3. Anchor damage				X
4. Cyclone damage				X
5. Pollution/Eutrophication			X	
6. Coral bleaching		X		
7. Coral pathogens/predators			X	
8. Freshwater runoff				
9. Siltation			X	
10. Fishing pressure		X		
11. Coral Cover	X			
BONUS/PENALTY POINTS	-20	-10	+10	+20
TOTALS	-20	-20	+30	+80

Each of 10 threat parameters and the coral cover category (11) is assigned various bonus or penalty points, utilizing a 4-tier system that reflects relative environmental damage: 1) excessive damage (-20 points), 2) moderate damage (-10 points), 3) light damage (+ 10 points), 4) no damage (+ 20 points). Coral cover is rated according to percentage of live hard coral as determined by 100 m line transects and is scored as follows: 1) < 26 %, 2) 26–50 %, 3) 51–75 %, 4) 76–100 %. In the example above, the resultant point total is 110. The maximum possible value of 220 (pristine reef with all parameters rated as category 4) is used to achieve equal weighting. The points total for each site is adjusted for equal weighting by multiplying it by 100 and dividing the result by 220. Therefore, for this example the adjusted figure is 50.

Calculation of Reef Condition Index – The sum of the adjusted total for each of the three main components described above. Each component contributes one third of the RCI, with a maximum score of 100 for each. Therefore, the top RCI for a totally pristine reef with maximum fish and coral diversity would be 300. Of course, this situation probably does not exist.

Interpretation of RCI values – The interpretative value of RCI will increase with each passing RAP. Thus far the complete data set contains 149 sites, 47 from the Togean-Banggai Islands, 57 from Milne Bay Province, and 45 from the Raja Ampat Islands. Table 1 provides a general guide to interpretation, based on the data accumulated thus far.

Table 1. Interpretation of RCI values based on 109 sites.

General reef condition	RCI value	No. sites	% of sites
Extraordinary	>243	5	3.36
Excellent	214–242	24	16.11
Good	198–213	40	26.85
Moderate	170–197	46	30.87
Poor	141–169	30	20.13
Very poor	<140	4	2.68

Coral cover

Data were collected at each site with the use of scuba-diving equipment. The main objective was to record the percentage of live scleractinian coral and other major substrates including dead coral, rubble, sand, soft corals, sponges, and algae. A 100-m measuring tape was used for substrate assessment in three separate depth zones at most sites, usually 4–6, 12–15 and 20–25 m. On several occasions there was only one or two transects were done. This was due to either insufficient depth, extremely strong current, or the presence of only one type of substrata (e.g. mud). Substrate type was recorded at 1 meter intervals along the tape measure, resulting in direct percentages of the various bottom types for each zone. For the purpose of calculating RCI, the average percentage of coral cover was used if more than one transect per site was involved.

Charismatic marine fauna

During the dive survey and while on route to and from sites, all participants noted any "charismatic" marine fauna. Charismatic marine fauna includes any marine species that appeal to non-scientists. Examples of these would include cetaceans, billfish, sea turtles, groupers (other large fish), rays, and sharks. These are listed under the sites where the animals were seen and also in a separate section where sightings occurred while the boat was underway.

Individual site descriptions

1. West end of Mansuar Island

Time: 0910 hours, dive duration 65 minutes; depth range 1–48.5 m; visibility 25–30m; temperature 28 °C; current none, relatively sheltered; visibility 25–30 m at all depths; *site description*: island fringing reef that is relatively sheltered; low diversity of *Acropora*; no dominant coral species noted across all zones; hard coral cover 34% at 4–6 m, 64% at 12–15 m and 52% at 20–25 m; average coral cover 50%; other substrata included patches of rubble and sand, soft corals, sponges, and tunicates; light fishing pressure with only moderate damage caused by fishing nets; one blacktip reef shark observed. RCI = 212.91.

2a. Cape Kri, Kri Island

Time: 1300 hours, dive duration 120 minutes; depth range 1–40 m; visibility 10 m at depth range of 0–19m and 15–20 m at depth > 20 m; temperature 28 °C; very strong current, at times high wave energy; slight turbidity; *site description*: fringing reef off of a small high island, very diverse amount of coral species with no dominant species noted from depth range of 5–20 m; dominant coral species *Acropora robusta*, *A. abrotanoides*, and *Pocillopora eydouxi* at 0–4 m depth range; hard coral cover 40% at 4–6 m, 19% at 12–15 m and 25% at 20–25 m; average coral cover 28%; other dominant substrata included rubble and sand followed by sponges, crustose coralline algae, and soft corals; presence of urchins, crinoids and brittle stars, *Drupella* observed; no threats or damage observed; two black tip reef sharks observed at 35 m depth. RCI = 231.28.

2b. Cape Kri Lagoon, Kri Island

Time: 0930 hours, dive duration 1:45minutes; depth range 1–26 m; visibility 15 m at depth of 1–19 m and 20 m at depth of 20 m; temperature 28 °C; current none; turbidity slight; *site description*: lagoon/sheltered bay with fringing reef outside (see site 2a) of small high island; dominant coral included strands of *Acropora* spp. at 18 m depth with foliose corals dominating along slope at approximately 15 m depth; seagrass bed with patches of sand and rubble found inshore shallows at approximate 2–4 m depth, hard coral cover 39% at 4–6 m and 47% at 12–15m, no transect done at 20–25m as mostly sand; average coral cover 43%, other dominant substrata was sand and rubble, strands of rubble covered with cyanobacteria and turf algae at 4–6 m depth; Other fauna included the giant clam, *Tridacna* spp. and the upside down jellyfish, *Cassiopea*, echinoderms (crinoids, urchins, sea cucumbers) and tunicates. Slight incidence of coral pathogens (white and black band disease). RCI = 170.56.

3. Mangrove Bay, South Gam Island

Time: 1600 hours, dive duration 90 minutes; depth range 1–24 m; visibility 5 m at 1–19 m depth and 15 m at 20 m depth; temperature 28 °C; current none; moderate to heavy turbidity; *site description*: fringing reef off of small high island with mangroves on shore; no transects done due to little coral coverage that ranged approximately from 10–20% at 1–19 m depth with sand found only at 20–25 m depth, other common biota included algae, sponges, echinoderms and tunicates; slight natural siltation noted on reef with little fishing pressure evident. RCI = 189.82.

4. North Kabui Bay, West Waigeo

Time: 1015 hours, dive duration 105 minutes; depth range 1–24 m; visibility 5 m at depth of 1–19 m and

15 m at depth of 20 m; temperature 28 °C; current none; turbidity moderate; *site description*: sheltered fringing reef off of undercut limestone island with mangroves on shore, reef slopes approximately 15 degrees levelling off at 5m; dominant coral was *Pachyseris* and *Galaxea* at 1–6 m depth; two transects were done at depth of 2–4 m and 4–6 m respectively as substrata consisted solely of mud past 6 m depth; hard coral cover 20% at 2–4 m and 72% at 4–6 m; average coral cover 46%; other biota included sponges, tunicates, and soft coral; slight freshwater run off and siltation; RCI = 141.44.

5. Gam – Waigeo Passage

Time: 1400 hours, dive duration 85 minutes; depth range 1 – 20 m; visibility 5 m at all depths (1–20 m); temperature 28 °C; current strong with low exposure to wave energy; turbidity moderate to heavy; *site description*: fringing reef in passage between two islands with mangroves along shore; dominant coral at 5–19 m depth was *Acropora florida*; other common coral was *Tubastrea* spp.; no transects done due to strong current; coral cover estimated at site to range from > 1% in some areas to 100% coral cover in strands of *A. florida*; other common fauna included tunicates, sponges, starfish, and soft coral; little to no fishing pressure, slight damage from pollution/eutrophication and presence of coral bleaching and pathogens; one manta ray, two cuttlefish, and schools of sweetlips seen. RCI = 122.39.

6. Pef Island

Time: 1600 hours, dive duration 60 minutes; depth range 1–42 m; visibility 15 m at 1–4 m depth, 20 m at 19 m depth and 25 m at depth > 20 m; temperature 28 °C; current none; *site description*: fringing reef off of small high island subject to some wave action, the inshore shallows have seagrass beds with patches of Sargasso; reef slopes at 30–35 ° angle to depth of 42 m where algae, mostly *Halimeda* spp.; and a few isolated boulders with small coral colonies occur; no dominant coral across depths, however high diversity of *Goniopora* spp., hard coral cover 53% at 4–6 m, 33% at 12–15 m, no transect was done at > 20 m because transect tape broke; average coral cover 43%; other common substrata/biota of transects included rubble, soft coral, and sponges; only light fishing pressure noted; other common fauna included lobsters, crinoids, sea stars (including *Acanthaster planci*), and urchins. RCI = 228.77.

7. Mios Kon Island

Time: 0900 hours, dive duration 90 minutes; depth range 3–35 m; visibility 10–15 m across depths; temperature 28 °C; current none; *site description*: platform reef off of small high island exposed to moderate wave energy, reef slopes at 25–30° angle; dominant coral at 4–8 m depth is *Acropora*,

Pavona, and *Diploastrea*, at greater depth patches of coral species mixed with soft coral and sponges interspersed with patches of sand and rubble, hard coral cover 11% at 4–6 m, 10% at 12–15 m and 19% at > 20 m; average coral cover 13.3%; other common substrata along transects included rubble and sand; other common biota noted included tunicates, sea stars, and sea cucumbers; slight siltation and fishing pressure with little incidence of coral bleaching noted on top surface of *Leptoria phrygia*. and *Goniastrea* sp. RCI = 222.44.

8. Mayalibit Bay, Waigeo Island

Time: 1200 hours, dive duration 105 minutes; depth range 1–21 m; visibility 5 m across depths; temperature 28°C; current strong; turbidity heavy; *site description*: fringing reef in sheltered bay off of small high island with mangroves, high siltation area; dominant substrata mixture of sand and silt with sponges and debris fields of dead coral and shells; dominant biota was sponges, *Phyllospongia lamellosa* with some *Dysidea* sp, hard coral cover 4% at 4–6m, 18% at 12–15m and 0% at > 20 m; average coral cover 7.3%; other common biota included algae, tunicates, sea stars, and crinoids; slight freshwater runoff with moderate siltation stress; RCI = 131.14

9. Mayalibit Passage, Waigeo Island

Time: 1530 hours, dive duration 90 minutes; depth range 1–20 m; visibility 5 m across depths; temperature 28°C; current strong; turbidity heavy; *site description*: fringing sheltered reef in wedge of small high island with mangroves, shallow reef flat with sea grass bed, reef slopes 30–45° angle that levels off at 20m with rubble, pebbles, and silt covering subtrata; no dominant coral species, but coral is exposed at low tide; hard coral cover 11% at 4–6 m and 21% at 12–15 m, no transect done at 20–25 m as no coral or other biota; average coral cover 16%; other common substrata of transects included sponges, soft coral, and turf algae; other common biota noted included tunicates, sea stars, and brittle stars; some damage was noted from explosive /cyanide fishing; slight fishing pressure; natural siltation evident with little freshwater runoff. RCI = 143.14

10. Pulau Dua, Wai Reefs

Time: 1000 hours, dive duration 90 minutes; depth range 1–23 m; visibility 15–20 m at 1–20 m depth and 20+ m at > 20 m depth; temperature 28 °C; wave exposure on south side; *site description*: platform reef with coral cays on platform, reef slope varies from 30–40° angle; dominant coral is *Acropora* spp. in 5 m depth with huge colonies of *Pavona* and *Porites* present; hard coral cover 40% at 4–6 m, 17% at 12–15 m and 17% at 20–25 m; average coral cover 24.7%; other common substrata/biota of transects included rubble, sand, and sponges, other common biota noted included algae, sea stars, sea cucumbers, brittle stars, tunicates, and crinoids; slight fishing pressure noted, evidence of blast fishing noted in other areas of this reef site not surveyed during this dive. RCI = 218.73.

11. North Wruwarez Island, (off of North Central Coast of Batanta Island)

Time: 1205 hours, dive duration 100 minutes; depth range 1–32 m; visibility 15 m average across depths; temperature 28 °C; sheltered, little to no wave exposure; *site description*: fringing reef off of small high island, reef slope varies from 30–40° angle; dominant coral of shallow depths (0–4 m) is *Montipora* spp. with deeper depths dominated by foliose coral; hard coral cover 56% at 4–6 m and 35% at 12–15 m, no transect done at 20–25 m depth as transect tape broke; average coral cover 45.5%; other common substrata of transects included rubble and sand; other common biota noted included algae, sponges, sea stars, sea cucumbers, brittle stars, tunicates, and crinoids; sea grass present inshore; moderate fishing pressure noted, some evidence of blast fishing, pollution/eutrophication, coral bleaching, freshwater runoff and siltation. RCI = 204.97.

12. South West Wruwarez Island, (off of North Central Coast of Batanta Island)

Time: 1500 hours, dive duration 90 minutes; depth range 1–32 m; visibility 5 m at 1–20 m depth and 12 m at > 20 m depth; temperature 28 °C; *site description*: fringing reef off of small high island within sheltered bay, mangroves present on shore, sand and then sea grass bed in shallows, reef slope varies from 30–40° angle; hard coral cover 36% at 4–6 m, 18% at 12–15 m, and 6% at 20–25 m; average coral cover 20%; other common substrata of transects included sand and rubble; other common biota noted included algae, sponges, sea stars, sea cucumbers, and tunicates; common algae was *Padina* sp. covered with cyanobacteria; coral *Porites cylindrica* noted to have parasitic flatworms, slight fishing pressure, siltation, freshwater run off, and pollution/eutrophication noted; close human population. RCI = 179.27.

13. Kri Island (site of dive camp)

Time: 1130 hours, dive duration 105 minutes; depth range 1–34 m; visibility 20 m average across depths; temperature 28 °C; exposed to strong current at times; *site description*: fringing reef with extensive reef flat filled with seagrass, soft corals, sponges, and patches of *Acropora*, reef slope ranges from 30–40° angle; slope of reef consists of mixed fields of rubble and sand with strands of live corals, mostly *Acropora* spp.; hard coral cover 41% at 4–6 m, 41% at 12–15 m, and 29% at 20–25 m; average coral cover 37%; other common substrata/biota of transects included rubble, sand, sponges, and dead coral; some

fishing pressure, anchor damage, siltation, and pollution/eutrophication noted; slight incidence of coral pathogens. RCI = 235.41.

14. Sardine Reef (Mios Kon between Koh Island)

Time: 1615 hours, dive duration 90 minutes; depth range 1–26 m; visibility 5 m at 1–19 m depth and 5–8 m at > 20 m depth; temperature 28 °C; exposed to strong current at times; *site description*: platform reef; mixed coral species across depths, hard coral cover 23% at 4–6 m, 24% at 12–15 m, and 16% at 20–25 m; average coral cover 21%; other common substrata/biota of transects included rubble, sand, and sponges; other common biota included algae and echinoderms; slight fishing pressure and coral pathogens; two sharks (tawny and wobbegong) seen. RCI = 210.17.

15. Near Dayang Island (off of Batanta Island)

Time: 940 hours, dive duration 90 minutes; depth range 1–41 m; visibility 10–15 m across depths; temperature 28 °C; protected, however exposed to strong current at times; *site description*: fringing reef with complex topography in channel between two islands; dominated by soft corals with some patches of foliose corals present at approximately 15m depth, sand and mud mixed with cyanobacteria at 20 m depth, black coral noted at depths > 20 m, hard coral cover 4% at 4–6 m, 7% at 12–15 m, and 8% at 20–25 m; average coral cover 6.3%; common substrata/biota of transects included soft coral, silt/mud, sand, and sponges; common sponges were *Dysidea* and *Phyllospongia*; other common biota included algae, some echinoderms (i.e. sea stars, crinoids, urchins), and tunicates; slight fishing pressure, net damage, siltation, freshwater runoff, and pollution/eutrophication. RCI = 210.17.

16. North West end of Batanta Island

Time: 1330 hours, dive duration 90 minutes; depth range 1–42 m; visibility 20+ m average across depths; temperature 28 °C; sheltered with exposure to current at times; *site description*: fringing reef off of small high island; dominant coral species was *Lobophyllia*, hard coral cover 50% at 4–6 m, 45% at 12–15 m, and 49% at 20–25 m; average coral cover 48%; other common substrata/biota of transects included sand, hydroids, and sponges; other common biota included sea grass, tunicates, and echinoderms (i.e. *Acanthaster plancii*, crinoids, brittle stars); slight fishing pressure and siltation with evidence of explosive/cyanide damage (one bomb scar); one wobbegong shark seen. RCI = 222.12.

17. West end of Wai Reef complex

Time: 1545 hours, dive duration 75 minutes; depth range 1–25 m; visibility 7 m average across depths; temperature 28 °C; exposed to strong current; *site description*: elongated ribbon-like platform reef; diverse coral species across depths, hard coral cover 54% at 4–6 m, 32% at 12–15 m, and 41% at 20–25 m; average coral cover 42.3%; other common substrata/biota of transects included soft coral, sponges, sand rubble; other biota present included algae, echinoderms (crinoids, urchins, sea-stars, and brittle stars), tunicates, and lobsters; slight fishing pressure noted. RCI = 209.84.

18. Melissa's Garden, North Fam Island
Time: 0930 hours, dive duration 120 minutes; depth range 1–40 m; visibility 15–20+ m average across depths; temperature 28°C; sheltered with exposure to current at times; *site description*: fringing reef surrounded by three small high islands; coral species diverse across depths; hard coral cover 53% at 4–6 m, 40% at 12–15 m, and 17% at 20–25 m; average coral cover 36.7%; other common substrata/biota of transects included sand, soft coral, and sponges (especially the boring sponge *Cliona* spp.); other biota present included seagrass, tunicates, and echinoderms (sea cucumbers, urchins, sea-stars, brittle stars); slight fishing pressure and coral pathogens evident. RCI = 223.34.

19. North Fam Island Lagoon
Time: 1145 hours, dive duration 75 minutes; depth range 1–32 m; visibility 10 m average across depths; temperature 28°C; *site description*: sheltered fringing reef in bay surrounded by small beehive islets around lagoon, mangroves present; diverse coral diversity with fields of *Oxypora glabra* dominant at approximately 20+ m depths; hard coral cover 50% at 4–6 m, 45% at 12–15 m, and 49% at 20–25 m; average coral cover 48%; other common substrata/biota of transects included sand, hydroids, and sponges; other common biota included sea grass, tunicates, and echinoderms (i.e. *Acanthaster planci*, crinoids, brittle stars); slight fishing pressure and some incidence of coral pathogens. RCI = 139.53.

20. North tip of North Fam Island
Time: 1545 hours, dive duration 75 minutes; depth range 1–25 m; visibility 35 m average across depths; temperature 28°C; periodic exposure to surge and small breaking waves; *site description*: fringing reef off of small high island with mangroves adjacent to site; hard coral cover 38% at 4–6 m, 26% at 12–15 m, and 23% at 20–25 m; average coral cover 29%; other common substrata/biota of transects included soft coral, rubble, sand (often mixed with silt), and sponges; other common biota included sea grass, tunicates, and echinoderms (crinoids, brittle stars, sea stars); slight fishing pressure; a shark and a hawksbill turtle seen (approximate length 0.5 m). RCI = 229.02.

21. Mike's Reef, SE Gam Island
Time: 0930 hours, dive duration 75 minutes; depth range 1–25 m; visibility 7 m average across depths; temperature 28°C; *site description*: platform reef off of rocky islet; strands of *Acropora florida* at 4 m depth, hard coral cover 19% at 4–6 m, 11% at 12–15 m, and 6% at 20–25 m; average coral cover 12%; other common substrata/biota of transects included soft coral, sand, sponges, and turf algae; other common biota included sea whips, tunicates, sea stars, sea cucumbers; slight fishing pressure with some damage from explosive fishing (bomb scar at 7 m depth) and fishing nets and line. RCI = 197.08.

22. Chicken Reef
Time: 1230 hours, dive duration 75 minutes; depth range 1–25 m; visibility 35 m average across depths; temperature 28°C; *site description*: platform reef with slope varying from 30°–40°; extensive rubble covered with turf algae and tunicates, strands of *Acropora florida* dominant at 4 m depth, hard coral cover 42% at 4–6 m, 26% at 12–15 m, and 5% at 20–25 m; average coral cover 24.3%; other common substrata/biota of transects included soft coral, rubble, sand, and dead coral; other common biota included tunicates, sponges, and echinoderms (crinoids, brittle stars, sea star, sea cucumbers), evidence of *Acanthaster plancii* present on reef although no individuals observed; slight fishing pressure and some incidence of coral pathogens; RCI = 200.09.

23. Besir Bay, Gam Island
Time: 1600 hours, dive duration 80 minutes; depth range 1–25 m; visibility 7 m average across depths; temperature 28°C; sheltered with little to no exposure to waves; *site description*: fringing reef under small high island inside cove like bay with mangroves, reef slopes at 25–30 degrees levelling off into sand/mud/silt fields; hard coral cover 57% at 4–6 m and 41% at 12–15 m with no transect done 20–25 m due to sand/mud/silt at depth > 20m; average coral cover 49%; other common substrata/biota of transects included algae, dead coral, rubble, and sand; other biota seen included tunicates, urchins, and sponges; light siltation, slight fishing pressure with some incidence of coral pathogens. RCI = 148.00.

24. Ambabee Island, South Fam Group
Time: 1000 hours, dive duration 105 minutes; depth range 1–42 m; visibility 20 m average across depths; temperature 28°C; *site description*: fringing reef off of small high islands, reef slopes 20° with extensive coral coverage on reef flat with diverse species of coral across depths; hard coral cover 57% at 4–6 m, 19% at 12–15 m and 4% at 20–25 m ; average coral cover 26.7%; other common substrata/biota of transects included soft coral, sand, rubble, and dead coral; other biota seen included tunicates,

sea cucumbers, and sea stars; slight fishing pressure with some incidence of coral pathogens. RCI = 200.46.

25. Southeast of Miosba Island, South Fam Island Group
Time: 1230 hours, dive duration 120 minutes; depth range 1–50 m; visibility 12 m average across depths; temperature 28°C; *site description*: platform reef sloping in 35–40°; *Acropora abrotanoides* dominat coral at 1–4 m depth, hard coral cover 11% at 4–6 m, 11% at 12–15 m and 5% at 20–25 m; average coral cover 9%; other common substrata/biota of transects included tunicates, rubble, sand, soft coral, and sponges; other common biota included algae, sea cucumbers, and sea stars; slight fishing pressure. RCI = 241.48.

26. Keruo Island, North Fam Group
Time: 1545 hours, dive duration 75 minutes; depth range 1–36 m; visibility 10 m at 1–4 m depth and 15 m at 5 – 20+ m depth; temperature 28°C; depending on area within site, exposure ranges from sheltered to exposed; *site description*: fringing reef 200 m from site 25 out to sea, more exposed with some areas sheltered; slope dominated by *Montipora florida* and *Acropora* spp., hard coral cover 48% at 4–6 m, 62% at 12–15 m and 50% at 20–25 m; average coral cover 53.3%; other common substrata/biota of transects included soft coral, rubble, and sand; other common biota seen included algae, brittle stars, tunicates, and sponges, presence of sea-stars and sea cucumbers noted; at reef edge schools of planktovirous fish; slight fishing pressure with blast fishing. RCI = 241.06.

27. Bay on Southwest Waigeo Island
Time: 0850 hours, dive duration 85 minutes; depth range 1–25 m; visibility 7 m average across depth; temperature 28°C; sheltered; *site description*: sheltered fringing reef in bay with mangroves onshore, hard coral cover 32% at 4–6 m, 26% at 12–15 m and 11% at 20–25 m; average coral cover 23%; other common substrata/biota of transects included rubble, silt, sand, and algae (mainly *Halimeda* and *Padina*); other common biota seen included brittle stars, tunicates, and sponges, presence of crinoids and urchins; slight siltation and eutrophication. RCI = 152.43.

28. Channel Between Waigeo and Kawe
Time: 1230 hours, dive duration 75 minutes; depth range 1–35 m; visibility 10 m across depth; temperature 28°C; strong current with some sheltered areas; *site description*: platform reef in channel off of three small high islands (beehive in shape), soft corals and crinoids dominant on substrata, hard coral cover 4% at 4–6 m, 4% at 12–15 m, and 8% at 20–25 m; average coral cover 5.3%; other common substrata/biota of transects included soft coral, algae, and sand; other biota present included brittle stars, tunicates, sea cucumbers, and sponges; high density of fish

with low diversity; no threats or damage noted; two manta rays seen. RCI = 177.07.

29. Alyui Bay, West Waigeo
Time: 1600 hours, dive duration 105 minutes; depth range 1–25 m; visibility 5–7 m across depth; temperature 28°C; sheltered; *site description*: reef far within sheltered bay with mangroves onshore, reef slopes approximately 20° with patches of hard and soft corals, sponges, and tunicates, strands of *Acropora* spp. dominate shallows at 1–2 m depth, hard coral cover 17% at 4–6 m, 28% at 12–15 m, and 5% at 20–25 m; average coral cover 16.6 %; other common substrata/biota of transects included soft coral, rubble, sand, and algae; other common biota seen included sponges, sea-stars, crinoids, and sea cucumbers; presence of lobsters, tunicates, and brittle stars noted; one *Acanthaster plancii* seen; slight fishing pressure, siltation, pollution/eutrophication with light incidence of coral pathogens. RCI = 180.64.

30. North end Kawe Island
Time: 0930 hours, dive duration 100 minutes; depth range 1–31 m; visibility 12 m across depth; temperature 28°C; medium to high wave energy; *site description*: fringing reef sloping approximately 10° peppered with a mix of sand, hard coral, soft coral, and rubble, mix of coral species, hard coral cover 19% at 4–6 m, 18% at 12–15 m, and 2% at 20–25 m; average coral cover 13%; other common substrata/biota of transects included rubble, algae, and sand; presence of crinoids, urchins, sea stars, sea cucumbers, brittle stars, and tunicates noted; slight fishing pressure. RCI = 213.81.

31. Equator Islands (east side)
Time: 1145 hours, dive duration 150 minutes; depth range 1–50 m; visibility 20–25 m across depth; temperature 28°C; *site description*: fringing reef, reef slopes ranges 20–35°, mix of *Heliphora* and *Porites cylindricus* dominant coral at approximately 3 m, large strands of foliose coral found at 20 m depth, hard coral cover 32% at 4–6 m, 19% at 12–15 m, and 20% at 20–25 m; average coral cover 23.7%; other common substrata/biota of transects included algae, sand, rubble, and in areas solid banks of rubble covered with turf algae; other biota seen included sponges, crustose coralline algae, and cyanobacteria; slight fishing pressure with blast fishing evident, slight siltation and eutrophicaiton, light incidence of coral bleaching, adjacent lagoon had mangroves. RCI = 230.39.

32. Equator Islands (west side)
Time: 1505 hours, dive duration 105 minutes; depth range 1–32 m; visibility 18–20 m across depth; temperature 28°C; sheltered with little wave action on upper reef; *site description*: fringing reef off of rocky islets, reef slope

ranges 35–40°, mix of coral species, hard coral cover 31% at 4–6 m, 10% at 12–15 m, and 11% at 20–25 m; average coral cover 17.3%; other common substrata/biota of transects included soft coral, sponges, rubble, sand, and algae; only slight fishing pressure noted. RCI = 204.33.

33. Alyui Bay entrance, Waigeo Island
Time: 0900 hours, dive duration 95 minutes; depth range 1–40 m; visibility 10 m across depth; temperature 28°C; subject to strong current in some areas; *site description*: fringing reef off of small high island, from approximately 28–35 m reef levels off where it undercuts island then continues incline with reef slope at 30° angel, hard coral cover 8% at 4–6 m, 9% at 12–15 m, and 3% at 20–25 m; average coral cover 6.7%; other common substrata/biota of transects included soft coral, sponges, and algae; presence of sea stars, crinoids, urchins, and tunicates noted; no evidence of threats or damage. One wobbegong shark seen. RCI = 217.64.

34. Alyui Bay entrance, Waigeo Island
Time: 1135 hours, dive duration 85 minutes; depth range 1–30 m; visibility 10–12 m across depth; temperature 28°C; *site description*: platform reef near tiny island at entrance of Alyui Bay, mix of coral species with abundance of sponge (likely to be *Ianthella* spp.), hard coral cover 16% at 4–6 m, 29% at 12–15 m, and 13% at 20–25 m; average coral cover 19.3%; other common substrata/biota of transects included soft coral, sponges, rubble and sand; other common biota included crinoids, urchins, and tunicates with some algae present; no evidence of threats or damage. RCI = 179.87.

35. Saripa Bay, Waigeo Island
Time: 1505 hours, dive duration 95 minutes; depth range 1–35 m; visibility 7 m at 1–4 m depth and 10 m at 5–>20 m depth; temperature 28°C; sheltered; *site description*: reef in sheltered bay with mangroves onshore, reef slopes at 40°angle, mix of coral species with beds of foliose corals at 15+ m, hard coral cover 46% at 4–6 m, 50% at 12–15 m, and 59% at 20–25 m; average coral cover 51.7%; other common substrata/biota of transects included algae, sponges, rubble, and sand; slight fishing pressure with siltation. RCI = 192.00.

36. Wayag Islands (east side)
Time: 0900 hours, dive duration 85 minutes; depth range 1–50 m; visibility 25 m across depth; temperature 28°C; exposed to high wave energy; *site description*: fringing reef off of small high island, reef slope varies from being sheer from 40–45 degree angle to gently sloping at approximately a 10 degree angle, hard coral cover 22% at 12–15 m, and 15% at 20–25 m with no transect done at 4–6 m depth due to heavy surge; average coral cover 18.5%;

other common substrata/biota of transects included soft coral, sand, and algae; presence of sponges, tunicates, sea stars, crinoids, sea cucumbers, brittle stars, and urchins; no threat or damage evident. RCI = 179.84.

37. Wayag Islands (west side)
Time: 1130 hours, dive duration 90 minutes; depth range 1–32 m; visibility 15 m across depth; temperature 28°C; *site description*: fringing reef near entrance to lagoon of Wayag Island, reef slopes at a 15–20°angle, the coral species, *Acropora palifera* dominates with strands of *Heliophora* in some areas, hard coral cover 49% at 4–6 m, 20% at 12–15 m, and no transect done at 20–25 m; average coral cover 34.5%; other common substrata/biota of transects included sponges, soft coral, and rubble; presence of algae, sea stars, sea cucumbers, and tunicates, some damage from blast fishing. RCI = 221.85.

38. Wayag Islands (inner lagoon)
Time: 1435 hours, dive duration 85 minutes; depth range 1–25 m; visibility 8 m across depth; temperature 28°C; sheltered; *site description*: reef in sheltered lagoon, reef slopes at a 30–40 ° angle, hard coral cover 25% at 4–6 m, 24% at 12–15 m, and 16% at 20–25 m; average coral cover 21.7%; other common substrata/biota of transects included sponges, sand, rubble, and algae; presence of sea stars, sea cucumbers, and tunicates; no damage or threats observed. RCI = 159.87.

39. Ju Island, Batang Pele Group
Time: 0915 hours, dive duration 90 minutes; depth range 1–38 m; visibility 12–15 m across depth; temperature 28°C; medium to light wave exposure; *site description*: fringing reef, shallows (1–2 m depth) consists mostly of rubble and rock mix with some *Acropora* spp., reef slopes at 35° angle that has large strands of soft coral, mix diversity of coral, hard coral cover 41% at 4–6 m, 35% at 12–15 m, and 16% at 20–25 m; average coral cover 30.7%; other common substrata/biota of transects included soft coral, rubble, algae, and dead coral; common biota tunicates and brittle stars, slight fishing pressure with some incidence of coral bleaching and pathogens. RCI = 239.49.

40. Batang Pele Island
Time: 1137 hours, dive duration 117 minutes; depth range 1–23 m; visibility 5 m across depth; temperature 28°C; sheltered; *site description*: sheltered fringing reef that slopes at 20° angle, sea grass and mangroves present, reef levels off at approximately 15 m depth into sand bottom with a few coral patches, hard coral cover 19% at 4–6 m and 13% at 12–15 m, and no transect done at 20–25 m due to only sand present; average coral cover 16%; other common substrata/biota of transects included algae

(mainly *Padina* and *Halimeda*), sponges, rubble, silt/sand; other biota included sea stars, brittle stars, and tunicates; slight fishing pressure with evidence of blast fishing, some incidence of coral bleaching and pathogens, some freshwater run off and siltation. RCI = 173.78.

41. Tamagui Island, Batang Pele Group

Time: 1445 hours, dive duration 75 minutes; depth range 1–32 m; visibility 8 m across depth; temperature 28˚C; light to medium wave exposure; *site description*: fringing reef with slope ranging from 35–40° angle, mangroves present inshore hard coral cover 40% at 4–6 m, 29% at 12–15 m, and 16% at 20–25 m; average coral cover 28.3%; other common substrata/biota of transects included sponges, sand, rubble, and algae; presence of sea stars, sea cucumbers, and tunicates; no damage or threats observed. RCI = 202.44.

42. Wofah Island, off Southwest Waigeo

Time: 0905 hours, dive duration 100 minutes; depth range 1–42 m; visibility 12–15 m across depth; temperature 28˚C; sheltered exposure; *site description*: fringing reef off of small high island with undercut limestone, from approximately 0.5 to 5 m depth dominated by the corals *Galaxae* spp., *Acropora bruggemaani*, and *A. palifera*, mixed coral diversity at deeper depths (6– >20 m), hard coral cover 45% at 4–6 m, 53% at 12–15 m, and 43% at 20–25 m; average coral cover 47%; other common substrata/biota of transects included rubble and sponges; other common biota included crinoids, sea cucumbers with some presence of brittle stars, sea grass, and tunicates; slight fishing pressure and light incidence of bleaching. RCI = 242.9.

43. Between Fwoyo and Yefnab Kecil Island

Time: 1200 hours, dive duration 90 minutes; depth range 1–23 m; visibility 18–20 m across depth; temperature 28 ˚C; protected with little to no exposure; *site description*:

platform reef with no cay or islet, slope ranging from 25–30° angle, coral present to depth of approximately 20 m with sand and rubble dominate > 20 m, coral species mixed across depth, hard coral cover 0% at 4–6 m, 27% at 12–15 m, and 45% at 20–25 m; average coral cover 24%; other common substrata of transects included sand, rubble, and sponges; presence of algae (mainly *Padina*), sea stars, sea cucumbers, sea urchins, and tunicates; slight fishing pressure and light incidence of bleaching. RCI = 208.03.

44. Yeben Kecil Island

Time: 1400 hours, dive duration 90 minutes; depth range 1–30 m; visibility 15 m across depth; temperature 28˚C; light wave exposure; *site description*: fringing reef off of small high island, reef slopes ranges from 20° angle to sheer spots up to 45° angle 40, cluster of mangrove trees present inshore, *Acropora florida* dominate at 6 m depth with patched of *Padina* and cyanobacteria, hard coral cover 37% at 4–6 m, 45% at 12–15 m, and 33% at 20–25 m; average coral cover 38.3%; other common substrata/biota of transects included sand, rubble, dead coral, soft coral, sponges, and algae; presence of sea stars, sea cucumbers, crinoids, tunicates, and sea grass; slight fishing pressure. RCI = 228.58.

Results

Reef condition

The hypothetical maximum RCI, as explained previously, is 300. During the current survey, values ranged between 123.49 and 258.82. The top 10 sites for reef condition are presented in Table 2. These are sites that have the best combination of coral and fish diversity, as well as being relatively free of damage and disease.

Table 2. Top 10 sites for general reef condition.

Site No.	Location	Fish species	Coral species	Cond. points	RCI
42	Wofah Island, off SW Waigeo	201	122	170	242.90
25	SE of Miosba I., S Fam Gp	281	83	170	241.48
26	Keruo Island, N Fam Group	263	79	190	241.06
39	Ju Island, Batang Pele Group	202	123	160	239.49
13	Kri Island dive camp	246	115	130	235.41
2a	Cape Kri, Kri Island	283	57	190	231.28
31	Equator Islands – E side	258	97	140	230.39
20	N tip of N Fam Island	214	92	180	229.02
6	Pef Island	209	94	180	228.77
44	Yeben Kecil Island	202	97	180	228.58

More than 50% of the reef sites surveyed during this expedition were ranked excellent to good. A summary of the frequency distribution for relative condition categories (see Methods section above for explanation) is provided in Table 3.

Table 3. Distribution of relative condition categories based RCI values for the reefs surveyed.

Relative condition	No. sites	% of sites
Extraordinary	0	0.00
Excellent	17	37.77
Good	10	22.22
Moderate	10	22.22
Poor	6	13.33
Very Poor	2	4.44

By the seven geographical regions surveyed, the sites of Pam Islands had the highest average RCI value of 212.48 (Table 4). The lowest average RCI value by geographical area was for sites within Mayalibit Bay. However, it is important to note that only two sites were surveyed within the bay.

Table 4. Average RCI values for major geographic areas surveyed.

Geographic area	No. sites	Avg. RCI
Fam Islands	6	212.48
Batang Pele to Pulau Yeben	5	210.46
Batanta – Wai Island	6	207.93
Kawe – Wayag Islands	7	198.17
Alyui Bay	6	194.25
Gam-Mansuar	13	193.10
Mayalibit Bay	2	137.14

By habitat type, fringing reefs had the highest average RCI values followed by platform reefs and sheltered bays (Table 5).

Table 5. Average RCI values for major habitat types surveyed.

Major habitat type	No. sites	Avg. RCI
Fringing reefs	23	214.58
Platform reefs	10	206.48
Sheltered bays	12	159.32

Coral cover

Coral cover ranged from a low of 5.3% at site 28, (the platform reef in channel between the islands of Waigeo and Kawe) and a high of 53.3% at site 26 (a fringing reef on Keruo Island, North Fam Island Group). Coral cover was generally higher at the shallower depth of 4–6 m. The highest number of coral species was 123 recorded at site 39 (a fringing reef off of Ju Island, Batang Pele Group). The lowest coral diversity occurred at site 5 (fringing reef in passage between the islands of Gam and Waigeo) with only 18 species of coral.

Coral bleaching,

Very low incidence of coral bleaching on reef sites (5, 7, 11, 31, 39, 40, 42, and 43). No serious or mass bleaching event noticed at any of the reef sites surveyed. Interestingly, locals recall no past major bleaching events on the reef areas studied.

Pathogens and predators

Coral pathogens observed on ten reef sites (5, 13, 14, 18, 22, 23, 24, 29, 39, and 40) with no major coral disease outbreaks noted. Common coral diseases noted included white and black band disease.

Very few coral predators observed. The coral-feeding mollusc *Drupella cornus* was only seen at reef site 2a (fringing reef off of Cape Kri Island). Crown-of-thorns starfish (*Acanthaster planci*) or evidence of its presence was seen at three sites (6, 16, and 19).

Siltation

Seventeen reef sites (2b, 3, 4, 7 – 9, 11 – 13, 15, 16, 23, 27, 29, 31, 35, and 40) surveyed had siltation. Seven sites (4, 8, 9, 11, 12, 15, and 40) had freshwater runoff as well. Nine of the sites with siltation stress were found in sheltered areas while seven of them were fringing reefs located close to shore. Only one platform reef, site 7, was found to have slight siltation stress. This reef was located in close proximity to a small high island. With the exception of site 8, all sites had slight siltation stress. Site 8 consisted of a reef located in Majabilt Bay of Waigeo Island and had moderate siltation stress.

Eutrophication/pollution

Slight evidence of eutrophication/pollution was observed at eight sites (5, 11-13, 15, 27, 29, and 31). All these sites were situated close to shore and included three sheltered bay reef sites (12, 27, and 29) and five fringing reefs sites (5, 11, 13, 15, and 31).

Fishing

Slight fishing pressure was evident at 32 sites (1, 3, 5–7, 10, 12–32, 35, 39, 40, 42–44) with moderate pressure observed at one site (11, fringing reef off of North Wruwarez Island). Evidence of various fishing methods was observed at a total of 11 sites. These included eight sites (10, 11, 16, 21, 26, 31, 37, and 40) for blast fishing, two sites (21) for the use of nets and one site (1) for the use of line.

Sightings of "charismatic" marine fauna

Sharks, rays, and turtles were the only charismatic marine fauna observed at the survey sites. A total of four black tip sharks (*Carcharhinus melanopterus*) were observed: one at sites 1 and 20 and two at site 2a. One tawny shark, *Nebrius ferrugineus*, was seen at site 14. A total of two wobbegong sharks were seen at sites 14 and 16. At site 28, two manta rays, *Manta birostris*, were observed while one was seen at site 5. One hawksbill turtle, *Erectmochelys imbricata*, was observed at site 20. Other marine fauna observed whilst underway were a pod (approximately 10 individuals) of short finned pilot whales, *Globicephala macrorhynchus*, and several (approximately seven individuals) of manta rays, *Manta birostris*.

Discussion

The average Reef Conditon Index (RCI) for the 45 sites surveyed in Raja Ampat is similar to the RCI value for the 47 reefs examined during the second Milne Bay survey in 2000. Both these average RCI values are significantly greater than the average RCI value for the 47 sites of the Togean Banggai survey (Figure 1). Like the Togean Banggai sites, no reef sites examined during the Raja Ampat survey had average RCI values ranked as extraordinary. However six reefs surveyed in Milne Bay (second expedition) had RCI values classified as extraordinary. A plausible reason for the average RCI values being greater at the Milne Bay and Raja Ampat sites compared to the Togean Banggai sites could be remoteness and population. The reef sites of the Togean Banggai Islands have a higher human population and are more accessible for marine resource exploitation, resulting in greater damage and reduced reef viability as indicated by their RCI values.

Remoteness and low population may also explain some of the patterns observed for the average RCI values within the seven major geographic areas surveyed during this expedition. The Fam Islands, wich have the highest average RCI value, are located southwest of Gam Island and southeast of Batanta Island (see map). The Fam Islands are uninhabited and located a fair distance from the larger and populated islands of Batanta, Gam, and West Waigeo.

In addition to location, habitat type also influences the condition of the reef or RCI value. The fringing and platform reef habitats had higher average RCI values than the sheltered bays. This can be expected as sheltered reef sites are usually subject to more natural siltation from land and low flushing rate or water exchange in the bays.

The most frequent observed stress or damage observed on reefs through out the survey region was fishing pressure. This is not surprising given that 90% of the inhabitants of Raja Ampat depend on marine resources for survival. Additionally, commercial fishing activities are also taking place throughout the region. Evidence of destructive fishing practices at 13.3% of the sites surveyed is particularly disturbing given the resulting damage and habitat loss. The use of destructive fishing methods involving cyanide and dynamite is reported by the local community to have increased in frequency over the past two years. Blast fishing is reported to be done by non-locals. The use of cyanide is reported to be by local villagers who are supplied the cyanide and squirt bottles by non-local merchant fishers of the live food fish trade (see chapter 6). Although blast and cyanide fishing is illegal, the activity occurs due to the difficulty with enforcement. As fish stocks are depleted throughout the western part of Indonesia and other Asian areas, fishers have ventured farther east to fish.

The low number of sharks seen during the survey most likely reflects the heavy shark finning activities taking place throughout Indonesia and globally. Sharks are heavily targeted for their fins, a key ingredient in the highly coveted shark fin soup that is considered a delicacy in Asia.

A potentially even greater threat to the coral reefs and other marine habitats of Raja Ampat is from poor land-use practices. Logging activities are occurring on the islands of Gam, Waigeo, and Batanta. The resulting deforestation from logging activity greatly compromises watershed integrity. Watersheds are a critical ecological link between land and sea. Poor land-use practices can result in water run-off carrying sediments, chemicals, pathogens (viruses and bacteria), and other pollutants to the reef. These compounds and substances introduced into the coastal waters can harm and kill coral and other reef organisms. Further, water quality and habitat conditions can be altered whereby a competitive advantage is provided to opportunistic non-reef building organisms (e.g. some species of algae and sponges). Mining is another land-based activity that is extremely harmful to the vitality of coral reefs. Nickel mining is proposed on Gag Island of the Raja Ampat and would have a devastating effect on the reefs if allowed to occur.

Other stressors and threats to reef vitality such as bleaching, coral diseases, and predators were observed sporadically on the reefs surveyed and were not extensive on the reefs where noted to occur. Locals when interviewed could not recall any mass bleaching event, widespread diseases or population outbreaks of coral predators on the reefs in the region. This is welcoming news given the increase and intensity of bleaching events such as the one that occurred north of the Raja Ampat Islands in Palau during the summer of 1998.

Conclusions

The majority of the reef sites surveyed during this expedition were in excellent to good reef condition. No damage or stress was noted on seven reef sites or 15.5% of the sites surveyed. Stress and damage was found on 84.5% of the reef sites surveyed. With the exception of three sites that had moderate levels of damage and stress, all other sites had minor damage and stress. The biggest threats to the reefs of the Raja Ampat Islands are from fishing intensity and use of destructive fishing practices. Current and proposed land-based activities such as logging and mining pose a serious threat to the reefs. Land use must be carefully planned and kept to a minimum if possible in order to keep the integrity of the watershed intact and to avoid excess siltation on the reefs. Results suggest that the reefs have incredible biodiversity and are under threat from anthropogenic activities. Marine and other natural resource exploitation of the islands will most likely increase unless proper conservation measures (e.g. education, management, enforcement, best practises) are taken and implemented as soon as possible. These activities should involve all stakeholders to preserve the incredible natural resources of the Raja Ampat Islands.

References

Bryant, D., L. Burke, J. McManus, and M. D. Spalding 1998. Reefs at Risk. World Resource Institute. 56 pp.

Spalding, M. D., C. Ravilious, and E. P. Green. 2001. World Atlas of Coral Reefs. Prepared at the UNEP World Conservation Monitoring Centre, University of California Press, Berkeley, U.S.A. 272-280.

Chapter 6

Exploitation of Marine Resources on the Raja Ampat Islands, Papua Province, Indonesia

Jabz Amarumollo and Muhammad Farid

Ringkasan

- Sekitar 90% penduduk di Kepulauan Raja Ampat hidup di daerah pesisir dan menggantungkan hidupnya pada sumberdaya laut.

- Masyarakat menggunakan beberapa metode penangkapan ikan, yaitu pukat/jaring, racun, busur dan panah, bahan peledak, dan jerat ikan (jerat tetap berukuran kecil dan besar).

- Sumberdaya laut juga dimanfaatkan untuk tujuan komersil. Namun hal itu bergantung pada keadaan permintaan pasar dan harga.

- Tingginya harga jual hasil laut dan manfaat lainnya dibandingkan pendapatan masyarakat tradisional memberikan insentif yang kuat untuk secara ilegal memanfaatkan sumberdaya laut secara berlebihan.

- Tingginya permintaan dan harga jual ikan hidup menyebabkan tingginya tekanan penangkapan ikan. Ikan target yang diperdagangkan adalah ikan karnivor dari famili Serranidae (kerapu) dan famili Labridae (ikan maming/napoleon).

- Permasalahan mendasar bagi masyarakat adalah desakan kebutuhan ekonomi; kurangnya pemahaman tentang hukum konservasi dan rendahnya kesadaran tentang perlunya konservasi.

- Banyak operasi pembalakan kayu dijumpai di daerah ini yang dapat mengancam ekosistem darat dan laut.

- Kepulauan Raja Ampat memiliki potensi untuk pengembangan wisata laut dan darat.

Summary

- Nearly 90% of the inhabitants of the Raja Ampat Islands dwell in coastal areas and depend on marine resources for survival.

- The communities use several fishing methods. These include nets, poisons, bows and arrows, explosives, and fish traps (both small and large stationary traps).

- Marine resources are also used for commercial purposes. However this depends on market demand and cost.

- High prices for marine goods and services relative to traditional community income provide a strong incentive for illegal overuse of marine resources.

- The high demand and market price of live food fish has resulted in high fishing pressure. Targeted fish for this trade include carnivorous fishes including fishes of the family serranidae (groupers locally referred to as kerapu or geropa) and of the family labridae (i.e. the Napoleon wrasse).

- Fundamental problems for the community are the urgent need for income, the lack of knowledge of conservation laws, and little awareness of conservation needs.

- Excessive logging operations are found in the area, which threaten terrestrial and marine ecosystems.

- The Raja Ampat Islands have potential for tourism development in both the marine and terrestrial sector.

Introduction

The Raja Ampat Archipelago is one of the island ranges in the western part of bird's head Papua (Indonesia). It covers approximately 6,962 km^2 or about 16.14% of the total area of Sorong Regency (43,127 km^2). The four large islands of Raja Ampat include Salawati, Batanta, Waigeo, and Misool. The archipelago is broken down into the five districts of Salawati, Samate, Misool, South Waigeo, and North Waigeo. These five districts have a total of 89 villages with a population of 48,707 and a population density of seven people per km^2 (Papua Province, census 2000).

Ethnic diversity in Raja Ampat is determined by "language families" or ethno-linguistic group. There are eight language family groups [1] in the islands. Most of the languages grouped into Austronesian. A few found in the western part of Salawati Island are classified as West Papuan.

There are five conservation areas (Nature Reserve or NR, known as a Cagar Alam in Indonesia) in Raja Ampat. These include West Waigeo NR, East Waigeo NR, West Batanta NR, North Salawati NR, South Misool NR, and the Raja Ampat Archipelago WS (Wildlife Sanctuary or WS, known as Suaka Margasatwa in Indonesia; BKSDA I, 1999). The conservation areas cover a total of 797,716 hectares consisting of 676,580 hectares of land and 121,136 hectares of sea. Raja Ampat was identified as a "*high-priority area*" for terrestrial and marine biological surveys by participants at the Irian Jaya Biodiversity Conservation Priority–Setting Workshop in Biak, Indonesia (1997), and at CI's Coral Reef Priority Setting Workshop in Townsville, Australia (1998). Conservation and economic activities within the region need to be balanced between long term sustainability of marine resources and the economy. Due to economic pressure, marine resource use, including destructive fishing practices, is increasing in Raja Ampat. This will likely result in more threats to marine biodiversity and impede conservation and management activities for sustained multiple use of the area by stakeholders.

Survey Method

Basic survey methods were used and consisted of Focus Group Discussion (FGD) and Direct Observation (Margoluis and Salafsky, 1998). An emphasis was placed on marine resource use by fishermen in the Raja Ampat Archipelago.

[1] The languages are "Maden, Palamul, Ma'ya" in Salaswati Island; "Legenyem, Waigeo, Biak" in Waigeo Island; "Kawe" in Batanta island; and "Matbat" in Misool Island (Based on Global Mapping Institute/Leontine Visser, 1998 in Workshop Map on Conservation Priority in Irian Jaya). Some languages have dialect similarity such as "Legenyem" in Waigeo and "Kawe" in Batanta island (Berry K., 2000).

Table 1. List of villages visited with location coordinates in the Raja Ampat Islands during the study.

Name of village	Name of The Island	Coordinates
Waiweser	North Batanta	00°45'04.6" S ; 130°46'59.1" E
Arefi	North Batanta	00°47'33.2" S ; 130°42'07.5" E
Yansaway	North Batanta	00°48'12.3" S ; 130°40'38.4" E
Marandan Weser	North Batanta	00°47'48.6" S ; 130°34'30.9" E
Saporkren	West Waigeo	00°26'14.7" S ; 130°39'58.1" E
Yenbeser	Gam	00°27'58.1" S ; 130°40'55.6" E
Friwen	Friwen	00°28'14.1" S ; 130°41'13.5" E
Yenbuba	Gam	00°34'16.5" S ; 130°39'14.8" E
Yenbekwan	Gam	00°34'29.3" S ; 130°37'50.0" E
Yenwaupnoor	Gam	00°31'41.1" S ; 130°39'14.8" E
Sawinggrai	Gam	00°32'02.2" S ; 130°34'51.3" E
Kapisawar	Gam	00°31'53.0" S ; 130°34'31.6" E
Arborek	Arborek	00°33'51.7" S ; 130°31'07.6" E
Wawiyai	West Waigeo	00°18'00.0" S ; 130°39'58.1" E
Kabui	Gam	00°27'17.1" S ; 130°33'06.0" E
Lopintol	West Waigeo	00°18'58.6" S ; 130°53'44.5" E
Waifoi	East Waigeo	00°06'04.4" S ; 130°42'49.0" E
Fam	Fam	00°40'13.8" S ; 130°17'44.7" E
Mutus	Batang Pele group	00°20'42.6" S ; 130°20'47.6" E
Miosmanggara	Batang Pele group	00°23'40.6" S ; 130°15'27.5" E
Manyaifuin	Batang Pele group	00°19'52.8" S ; 130°13'00.9" E
Selpele	West Waigeo	00°12'15.8" S ; 130°13'25.2" E
Salio	West Waigeo	00°07'25.4" S ; 130°17'31.0" E

Table 2. Names of fishing companies that use marine resources of Raja Ampat with a listing of commodity (taxonomic family name is given in parentheses below the common name) and production rate (tons/year) for 1999. (PT. denotes limited company while CV. denotes a small company that is usually family owned.)

No.	Company Names	Commodity	Production (Tons/year)
1.	PT. Usaha Mina	Mackerels/Tuna (Scombridae)	6,063.4
2.	PT. Ramoi	Mackerels/Tuna (Scombridae)	3,559.9
3.	PT. Citra Raja Ampat Canning	Mackerels/Tuna (Scombridae)	2,199.5
4.	PT. Keselamatan Cinta Bahari	Mackerels/Tuna (Scombridae)	2,092.4
5.	PT. Ponco Susetyo Sakti	Ikan Kerapu (Serranidae)	3.0
6.	PT. Citra Karya Permai	Ikan Kerapu (Serranidae)	7.3
7.	PT. Hasuda Mina Bahari	Ikan Kerapu (Serranidae)	4.5
8.	CV. Winka	Ikan Kerapu (Serranidae)	34.6
			13,964.6
9.	PT. Arta Samudra	pearl	10.20
10.	PT. Megapura Aru Mutiara	pearl	
11.	PT. Yellu Mutiara	pearl	
Total			**13,974.8**

Source: Fishery Departement of Sorong, 2000

A total of 23 villages (Table 1) consisting of four villages on North Batanta Island and 19 villages on South Waigeo Island were visited and surveyed from 26 March to 11 April 2001. The combined population of these 23 villages is 5,726, ranging from 98 in Arborek village to 785 in Fam village. Data collected included people's marine resource use activities and socio-economic views (e.g. type of marine biota captured, tools and techniques used, threats and conservation efforts and, marine management and conservation knowledge).

Results and discussion on marine resource use in Raja Ampat is generally divided into two socio-economic patterns, namely traditional and modern sectors. The majority of marine resource use falls into the modern sector, as Raja Ampat becomes a fishing base (for traditional fisherman as well as commercial companies). The name of fishery companies located in Sorong are given in Table 2 along with commodity and production rate.

Total fishery production was 40,828.72 tons per year for 1999 in Sorong Regency. Of this total, 13,964.6 tons or about 34.20% was composed of tuna and a large number of carnivorous fishes. Also that year (1999), pearl production originating from Raja Ampat was 10.20 tons in Sorong Regency. Internationally, the products were exported to America, Hong Kong, Japan, China, and Europe. Domestically the pearl products are sent and marketed in Jakarta, Surabaya Makasar, Bali, and Jayapura, while local distribution takes place in the vicinity of Sorong.

Nearly 90% of the population dwells along the coast and depends on marine resources for survival. Traditional fishing methods are used and usually take place in shallow water areas within 200 m depth. Various simple methods for catching fish are used by the inhabitants. Generally, marine resources are obtained in two ways. These include hunting in order to catch fish and mariculture [2].

Fishing

Surface fish, coral reef fish, pelagic fish, shells, and other molluscs are caught by local inhabitants in shallow areas, usually less than 200 meters from shore. Large-scale fish companies use modern fishing techniques in the open ocean. Inhabitants generally use marine resources for family consumption. However some fish for commercial purposes. The amount of fishing for commercial purposes varies depending on market demand. Presently, there is a high demand for Napoleon Wrasse, Crustaceans (lobster), and shark. Other organisms such as holothuroids (sea cucumbers), *Trochus niloticus* (trochus), and *Pinctada* spp. (*Japing-japing* or pearl oysters) are also in demand on the market.

Napoleon Wrasse and serranids (sea basses and groupers) are caught alive for commercial purposes. These fishes have a high market price and are easy to capture. The fish are caught alive by stunning them with poisonous chemicals such as cyanide. The practice originated outside of Raja Ampat and was introduced to the local communities who are now using it. Facilities with underwater holding pens to store the cyanide-caught live fish are found in the villages of Fam, Miosmanggara, Mutus, Yembekwan, Arefi, and Arborek. The fish are usually loaded and shipped from the villages of Mutus and Miosmanggara, where there are loading ports. The various techniques for catching fish are listed and described in Table 3.

[2] People of Wawiyai cultivated sea slugs in 30-metre square areas using a traditional method called "warin purin." The slugs were caught at night using a kerosene lantern called a lobe and raised in the places provided. Five hundred individuals were raised. They failed to harvest from their activities, however, because people abandoned the process and were lacking in knowledge of cultivation techniques.

Table 3. Fishing methods used in the Raja Ampat Islands. The tools used for the fishing methods along with the marine biota targeted are given with remarks.

Tools used to capture marine biota	Descriptions	Targeted biota	Remarks
Indigenous subsistence level fishing methods			
Hook and line	A line is tied with one to many fish hooks that vary in size.	Pelagic and demersal fish (e.g. tuna, groupers, mackerels, jacks and, coral fish).	Technique involves setting a long line or a drift long line. Locally called "*throwing lot*" and "*throwing hook.*"
Spears and bows	Iron spears are tied on long wood poles about 2-3 m in length. Spears are thrown by hand from the surface or underwater while diving. If used underwater, a rubber sling shot provides propulsion.	Pelagic and demersal fish.	Local method is called *lobe* and is done at night using row boats and kerosene pressure lanterns. The spears used either have many arrows (*Kalawae*) or one arrow (acu). Method sometimes used without snorkeling or diving equipment.
Legal commercial fishing methods			
Hook	Techniques are the same as mentioned for hook and line except the hook(s) are attached to a rope. Commonly used as method of fishing from the surface on boats while underway or trowling.	Pelagic fish such as *kembung* (mackerel) tuna, ray fish, oil sardine; demersal fish such as *bubara* (pompano), shark, *kurisi* (*Nemipterus nematophorous*), *bulanak* (*Valamugil* sp.), and serranids.	Motorboats are used at speeds of 5-25 knots. Number ropes deployed off boat stern depends on need; usually 5-10 ropes are used.
Nets	Cast into the water above from surface or set underwater.	Pelagic fish such as *kembung* (striped mackerel), shark, *lalosi* (fusilers), samandar, and *baronang*(*Siganus* spp.), and demersal fish such as serranids and coral reef fish.	Used by non-locals to catch schools of fish.
Large fish trap (Sero)	The trap is made of bamboo and is triangular in shape. It is used in shallow reef areas and set to work with the rise and fall of the tide.	Pelagic fish such as *bubara* (pompano), *lalosi* (fusilers), and other coral reef fish.	This is not widely used but is an effective method.
Fish trap (bagan)	This type of trap is made of net and bamboo. A light is used to attract fish.	Pelagic such as teri (a kind of small baitfish), *kembung* (striped makerel), *lalosi* (fusilers), etc.	This method causes over-fishing.
Illegal commercial fishing methods			
Poison	This method uses poisonous chemicals such as cyanide and potassium chloride. It is used to stun the fish and often leads to death of the fish and the coral.	Generally used to catch carnivorous bony fishes and serranids, demersal fish, and lobster, etc.	From squirt bottles, poison is applied directly into the water near the targeted fish (e.g. coral crevices where fish hide). Compressor or hookah system is used to assist divers.
Bomb/explosive material	Explosive materials used are urea-based fertilizers, matches sulfur, poisonous bullets that are tightly packed into bottles with wicks for fuse.	Schooling pelagic fish such as *bubara* (pompano), *lalosi* (fusilers), *kembung* (striped mackerel), etc.	Fishermen from Sorong commonly use this method; however local communities are also using it.

Prices

Prices quoted are for the selling of the fish by the fishermen to the fish merchants. Seabasses and groupers (serranids) and the Napoleon Wrasse (*Cheilinus undulatus*) cost about Rp. 3,000 per kilogram Serranids can sell for Rp. 80,000 per kilogram while Naploean wrasse can sell for Rp. 130,000. Only seven Napolean Wrasse (*Cheinilus undulatus*) were seen during the survey, which may indicate a decline in their population numbers. Shipping capacity of fish was 100 – 5000 kilograms per export load from Miosmanggara village and 200 – 400 kilograms per export load in Mutus village. Exporting of fish usually takes place every two to three months. In 1999, two shipments took place from Misomanggara while three took place from Mutus. Groupers (known locally as *Tongseng Is* and *Saisseng Is* or scientifically as *Plectropomus leopardus* and *P. areolatus*) are transported from Bitung, Ternate, and Sorong to Hong Kong by grouper tradesmen.

Commercially, marine goods are sold locally in Sorong by merchants from the villages of Makasar and Buton. Average prices of these commodities are as follows:

Teripang

Sea cucumber (*teripang*) are sold for about Rp. 15,000–80,000 per kilogram and depending on species can sell for Rp. 130,000 per kg. The amount of sea cucumbers exported ranges from 5–100 kilograms per month. Income for fishers selling sea cucumbers to exporters ranges from Rp. 50,000–600,000 every month depending on kind and quality of sea cucumbers sold. Collectors go to buy sea cucumbers from the fishermen either once a month or once every three months. The irregularity in the visits from the collectors can cause a disruption in income for fishermen selling in Sorong.

Lobster

Lobster sells for Rp. 55,000 to 60,000 in Sorong. The amount of lobsters exported ranges from seven kilo grams to 20 kilo grams per month, which is equivalent to a cash payment of approximately Rp. 400,000–1,000,000 per month depending on the amount of fishing activity. Local people only sell to collectors for fear of being arrested, as town people say lobsters are protected.

Japing-japing (Pinctada spp. or pearl oysters)

Depending on market demand, *japing-japing* ranges in price from Rp. 3,500 to 5,000 per kilogram in Sorong. The amount of japing-japing exported ranges from 5–20 kg per month or a cash equivalent of approximately Rp. 3000–50,000- per month. People view the selling of oysters as a secondary activity because of selling difficulty in Sorong.

Table 4. Activities that threaten marine resources in the Raja Ampat Islands. The village where the activity takes place is given by number as listed in Table 1, along with the impact and comments.

Threat Type	Village Location	Impact	Comments
Explosive/blast fishing	1-5, 12, 18-23	Damages and kills coral and pelagic fish.	Agents are usually from Sorong and use long boats.
Cyanide fishing	2, 3, 6, 9, 18-22	Kills coral and demersal fish.	Intended to stun marine biota (serranids and lobsters) to catch for live food fish trade; fishers often aided by compressors.
Over fishing	2, 3, 6, 9, 11-14, 16, 18-23	Decrease in populations of serranids, teripang (sea cucmber), lola (trochus), and teri.	Fishing intensity is high depending on market demand.
Government planned buildings	5, 14, 15	Clearing land for building and construction purposes. Results in coastal erosion and sedimentation to marine ecosystems.	Developmental planning for Raja Ampat Regency is scheduled for the future by the Local Planning Board in Sorong.
Logging[3]	5, 11-17, 23	Clear cutting of forests causes coastal erosion and sedimentation on the coral reefs. Also a source of "social conflict."	High intensity logging done by companies and local people.

[3] Not only in Waigeo Island, Batanta Island in Yenanas village, and Salawati Island, but also the activity has been done in villages of Kaliam, Solol, Kapatlap, Samate, Kalobo, and Waijan.

Shark fin

Shark fin ranges in price from Rp. 50,000 to 80,000 per kg per month depending on supply. Even though the shark fin trade can contribute substantial income for the fishermen, few engage in the activity due to the high cost of tools, the large time and energy commitment, and the need for special skills. The selling of shark fins to exporters is usually done locally in the villages, then transported to Sorong.

Ikan teri

These small silvery fishes (*ikan teri*) usually include members of the families Engraulidae (anchovies) and Clupeidae (herrings). Depending on type and quality of *ikan teri*, prices can ranges from Rp. 3,000 to 7,000 per kg. The amount of *ikan teri* that is supplied to consumers per month varies from 1/2 to 1 ton or a cash equivalent of Rp. 200,000 – 3,000,000 depending on availability of tools in such villages as Wawiyai. Fishing for *ikan teri* is limited in practice due to the high cost of boats and supplies used (Rp. 3,000,000 – 5,000,000 per equipped boat or bagang). In addition to the bagang, nets, and kerosene lamps are needed for *ikan teri* fishing. *Ikan teri* is sold locally to collectors who visit on a regular monthly basis.

Ikan asin

Salted fish (*ikan asin*) consists of *tenggiri* or makerel (*Scromberomorus spp.*) and ikan batu (demersal and coral fishes). Depending on the quality, *tenggiri* is sold for Rp. 10,000 – 12,000 per kg in the villages and Rp. 15,000 – 20,000 per kg in Sorong. The market price of *ikan batu* ranges from Rp. 5,000 – 6,000 per kg in the local villages and Rp. 6,000 – 8,500 per kg in Sorong. Local collectors also buy fresh fish from the communities at a price of Rp. 3,000 per kg for *tenggiri* and Rp. 2,000 per kg of coral fish. Production capacities for a mixture of both types range from 25 to 500 kg per month depending on catching intensities and seasons or cash payment ranging from Rp. 45,000 to 800,000 per month based on kinds and qualities. The *ikan asin* trade is commonly done by Raja Ampat people because it demands little money and follows very simple methods from fishing to salting to marketing.

Pearl

Pearls locally cost Rp. 7,000 each. Maximum production is 2,000 pearls per month or a cash equivalent of Rp. 14,000,000. This biota is directly taken from nature by divers with or without the use of a compressor. This activity occurs only in Lopintol village where pearl entrepreneurs have supplied the tools and take care of marketing.

Other marine biota

Other marine biota used by the local people include mangroves for house construction, molluscs, crabs, and shrimps for food. This biota is limited solely to family use and has not been commercialized, although quantitatively there is potential.

Threats to marine resources of Raja Ampat

Various direct and indirect threats to marine resources were identified during the survey and are listed in Table 4.

Several factors contribute to the use of illegal fishing methods. First, local people sell their fish products at low prices; however goods and services are purchased at high prices. This results in a deficit in family finance and an increase for basic needs.

Second, some regions are classified as of low economic standard. These regions continuously exploit marine and coastal resources illegally, as it is the only way for them to meet their needs. Of the 23 villages surveyed in Raja Ampat, 96.4% or 898 families in 19 villages of South Waigeo were categorized as pre-prosperous group. Twenty-three villages are still actively using illegally fishing methods for both family and commercial purposes.

Conservation efforts

The forest, marine, and coastal ecosystems of the Raja Ampat Islands are biologically rich with abundant natural resources. The government has allocated 797.72 hectares of these ecosystems as a nature conservation area. This area can be divided by ecosystem into 676.58 hectares (84.8%) of forest and 121.14 hectares (15.2%) of coastal and marine habitat spanning over five districts in Raja Ampat.

The following stipulations [4] were made by the local government of Sorong Regency in an effort to protect marine resources:

1. Limit number of Teri catching licenses that use large fish traps. These traps were responsible for an estimated 95% of the over fishing or an equivalent of 1.15 tons per year.

2. Prohibited use of compressors (hookah) and scuba to catch marine biota such as large carnivorous fishes, serranids, Napoleon wrasse, and lobsters. This activity usually involved the use of poisons such as potassium cyanide.

3. Limit number of large carnivorous fishes (including serranids and Napoleon wrasses) and lobsters that can be caught.

---

[4] The stipulations were made in consideration of the negative impact caused by outsiders (fishermen) extracting marine resources (Local Fishery Office Sorong, 2000; Yearly Report by Fishery Office Sorong, 1999).

Table 5. Marine resource use of Miosmanggara village. Commodity, production, price, and remarks are given.

Commodity	Production (kg/month)	Price (Rp)*	Remarks
Carnivorous fishes including serranids	100–500	5,000–60,000	Consists of various kinds and classes
Teripang (sea cucumbers)	100–500	60,000–80,000	Consists of various kinds and classes
Salty fish	100–500	5,000–10,000	Tenggiri and coral fish
Lobster	10	15,000–50,000	Consists of various kinds and classes
Lola (trochus)	5–20	15,000	Consists of several kinds
Shark fin	5–10	50,000–350,000	Consists of various kinds and classes

Ranges of price are based on reports from local fishermen.

4. Issue permits for catching of carnivorous fishes including serranids, Napoleon wrasses, and lobsters under conditions below:

   a. Collectors consist of members from the local community. The permit holders can be outsiders who reside in the local village.
   b. Local fishermen are informed of their fishing activities.
   c. Create good cooperation with fishermen (PIR patterns), system where entrepreneur buys tools for local fishermen. The catch from fishing activities is split into equal shares, one for the fisherman and one for the entrepreneur. When the tools are fully paid by the fishing activities, the fisherman owns the tolls.

Only two villages, Arefi and Yansaway on Batanta Island, expressed interest in re-instituting the traditional conservation policy known locally as "*sasi gereja.*" These villages believe "*sasi gereja*" is effective over the local community as it involves traditional law as well as the church. Due to the considerable decline in marine resources and environmental quality, traditional conservation policy will be re-instituted on Way Island. Way Island is the site visited by these two villages as this island is rich in marine resources (mainly fish, sea cucumber, clams, and lobsters).

In order to implement "*sasi gereja*," a church service is conducted whereby alms are presented to announce "*sasi*" or a moratorium on fishing activities. A service is then held after an extended period of time based on the assumption that the natural resources have had enough time to recover. Penalties are given to those that do not follow the "*sasi gereja.*"

Traditional laws are not effectively used by local communities to solve problems regarding natural resource use. Instead local communities wait for solutions to the problems by local government officials in the Districts. During our survey one local person, Taher Arfan,[5] said that problems occur as a result of natural resource use by outsiders even though the Local Traditional Board still functions (Personal communication, 2001).

In general, local people's knowledge or understanding about natural resources conservation is very limited. Often this results in protected marine biota, such as turtle, lola, batu laga, and Napoleon wrasses being exploited. There are little to no conservation awareness activities due to the geographical isolation and resource limitations (e.g. educational tools and trained personnel) of the area.

Community's perspective

Several issues are very important to the people in Raja Ampat regarding natural resource use of forest, marine, and coastal ecosystems. The people interviewed in Miosmanggara village were critical, honest, and at times contradictory regarding the use of marine resources and the accompanying socio-economics.

Miosmanggara is a village with a population of 220 and a density of 0.04 people per km^2 located within the Batang Pele Island Group. Approximately 84.6% of the population are categorized as pre-prosperous or at poverty level. In this village, people are widely known to actively engage in destructive fishing methods such as cyanide and blast fishing. Commodities associated with the villagers' marine resource use, production, and price are given in Table 5.

The social economic data gives the impression that the village's standard of living is good when in fact their monthly cash income ranges from 50,000 to 250,000 rps per month. However this contradicts the entrepreneurs who state that they spend 13,000,000 to 15,000,000 to buy live fish alone not including any other associated business expenses (e.g. gasoline for transport to Miosmanggara village, staff of live fish merchant companies).

[5] The head of Raja Ampat Indigenous People Board "Kalanafat."

According to the villagers, they can get enough income to support themselves. They therefore said: *"We can stop the use of illegal catching methods, if we are given compensation or other alternative that allows us to meet our needs now. We know that what we are doing is illegal; however as long as there are no alternatives, we will continue. After we eat enough then punish us. If there are any alternatives such as job opportunities with good salary then we will let entrepreneurs (either loggers or tourism businesses) operate here. We will give them a low rental price for use of our land. The most important issue to us is for the entrepreneurs to provide skill training for our youths. That way our young people can work in the companies that operate here."*

Threats to the forest ecosystems are caused by large-scale logging activities by a conglomerate of companies[6] or by the community. Some villages (Arborek, Sawinggrai, Kapisawar, Wawuyai, and Sapokren) have suffered the consequences of these activities as they often create conflicts, especially social ones involving traditional land ownership. Often a logging businessman makes an agreement with the village head (locally called *kepala desa*) to allow logging activities in the area. In exchange, the businessmen promises to build a community center, church, or house for the village. Sometimes the full details of the agreement are not revealed to all the villagers, leading to confusion. Reports of the villagers not receiving the promised deliverable (e.g. building) from the businessman occur. The exact details of these events are not clear as the agreement is often not fully disclosed to all stakeholders.

Tourism development in Raja Ampat

Raja Ampat has amazing ecosystem diversity. Within these ecosystems (forest, marine, and coastal) lies incredible biodiversity, some of which has yet to be documented. For example, on the land many plant species are endemic. The forests contain endemic and rare bird species such as Waigeo Brush Turkey (*Aepypodius bruijnii*), Red Bird-of-Paradise (*Paradisaea rubra*), Wilson's Bird-of-Paradise (*Diphyllodess republica*), Northern Cassowary (*Casuarius unappendiculatus*), and the Western Crowned Pigeon (*Goura cristata*).

In addition to the forests, the marine ecosystems of Raja Ampat have remarkable potential to draw tourists. Biologically, the marine biota such as corals, molluscs, and fishes are phenomenal. Ecological tourism

development on the Raja Ampat Islands is very poor in comparison to other places such as Bunaken Island in North Sulawesi, even though Raja Ampat has a more diverse fauna. Furthermore, many potential sites for diving and snorkeling were identified by the Marine RAP team. These included dive sites on the islands of Kri and Wei.

The local office of tourism in Sorong Regency revealed that only one company, Irian Diving, has been legally operating as a dive tourist center. This company continues to develop and promote tourism in the Raja Ampat Islands. The local office of tourism reported that there were 184 tourists in 2000 with Raja Ampat as their destination. While the company,[7] Irian Diving reported that visitors have been progressively increasing. For example, in 1999 and 2000 respective totals of 66 and 201 tourists were reported. Most of the visitors were Americans who stayed at the center for one to three weeks. The local governmental planning board of Sorong Regency in collaboration with local office of Tourism in Sorong plans to develop tourism with an emphasis on nautical tourism for the Raja Ampat Islands. It is hoped that the results from the Marine RAP and the associated follow up activities will help contribute to tourism development (Soedirman[8], Personal communication).

Reference

Berry K., Community Report, PT. Cendana Indopearls, Selpele and Salio.

BPS Irian Jaya, 2000. Irian Jaya Dalam Angka. Balai Pusat Statistik Irian Jaya.

Margoluis, R. and N. Salafsky, 1999. Measure of Success. Island Press, Washington DC.

Dinas Perikanan Sorong, 2000. Annual Report of Fishery Department Sorong Regency (unpublished)

---

[6] Timber industry actively operates 9 companies, while the people operate 8 companies.

[7] PT Irian Diving, a diving tourism service company operating in Raja Ampat and based on Kri Island, South Waigeo District.

[8] Section head of physics and infrastructure of Local Planning Board, Sorong Regency.

Pef Island near Site 6. (R. Steene)

Shallow reef corals, Pam Islands. (R. Steene)

Base camp at Kri Island. (G. Allen)

Eviota raja, a new goby species. (G. Allen)

Orbicular batfish, *Platax orbicularis*. (G. Allen)

Mayalibit Passage, Waigeo Island. (G. Allen)

Crinoids on sponge. (R. Steene)

Red soft coral, *Dendronephthya* sp. (R. Steene)

Beehive islets at the Wayag Islands. (G. Allen)

Giant clam, *Tridacna crocea* (R. Steene)

2001 Raja Ampats Marine RAP Science Team.
Back row (L to R): J. Veron, F. Wells, S. McKenna, R. Steene, and D. Fenner.
Front row (L to R): La Tanda, M. Farid, G. Allen, P. Boli, and J. Amarumollo. (Max Amer)

Appendices

APPENDIX 1

Checklist of corals of eastern Indonesia and the Raja Ampat Islands

J.E.N. Veron

Zooxanthellate Scleractinia	Recorded in eastern Indonesia (Veron, 2000)[1]	Recorded in Raja Ampat Islands
Family Astrocoeniidae		
Stylocoeniella armata (Ehrenberg, 1834)	X	X
Stylocoeniella guentheri Bassett-Smith, 1890	X	X
Madracis kirbyi Veron and Pichon, 1976	X	
Palauastrea ramosa Yabe and Sugiyama, 1941	X	X
Family Pocilloporidae		
Pocillopora damicornis (Linnaeus, 1758)	X	X
Pocillopora danae Verrill, 1864	X	X
Pocillopora eydouxi Milne Edwards and Haime, 1860	X	X
Pocillopora kelleheri Veron, 2000	X	X
Pocillopora meandrina Dana, 1846	X	X
Pocillopora verrucosa (Ellis and Solander, 1786)	X	X
Pocillopora woodjonesi Vaughan, 1918	X	X
Seriatopora aculeata Quelch, 1886	X	X
Seriatopora caliendrum Ehrenberg, 1834	X	X
Seriatopora dendritica Veron, 2000	X	X
Seriatopora guttatus Veron, 2000	X	X
Seriatopora hystrix Dana, 1846	X	X
Seriatopora stellata Quelch, 1886		X
Stylophora pistillata Esper, 1797	X	X
Stylophora subseriata (Ehrenberg, 1834)	X	X
Family Acroporidae		
Montipora aequituberculata Bernard, 1897	X	X
Montipora altisepta Nemenzo, 1967	X	X
Montipora angulata (Lamarck, 1816)		X
Montipora australiensis Bernard, 1897	X	X
Montipora cactus Bernard, 1897		X
Montipora calcarea Bernard, 1897	X	
Montipora caliculata (Dana, 1846)	X	X
Montipora capitata Dana, 1846	X	X
Montipora capricornis Veron, 1985	X	X
Montipora cebuensis Nemenzo, 1976	X	X
Montipora cocosensis Vaughan, 1918		X
Montipora confusa Nemenzo, 1967		X
Montipora corbettensis Veron and Wallace, 1984	X	
Montipora crassituberculata Bernard, 1897	X	

Zooxanthellate Scleractinia	Recorded in eastern Indonesia (Veron, 2000)[1]	Recorded in Raja Ampat Islands
Montipora danae (Milne Edwards and Haime, 1851)	X	X
Montipora delicatula Veron, 2000		X
Montipota digitata (Dana, 1846)	X	X
Montipora efflorescens Bernard, 1897	X	X
Montipora effusa Dana, 1846	X	
Montipora florida Nemenzo, 1967		X
Montipora floweri Wells, 1954	X	X
Montipora foliosa (Pallas, 1766)	X	X
Montipora foveolata (Dana, 1846)	X	
Montipora friabilis Bernard, 1897	X	X
Montipora gaimardi Bernard, 1897	X	X
Montipora grisea Bernard, 1897	X	X
Montipora hirsuta Nemenzo, 1967		X
Montipora hispida (Dana, 1846)	X	X
Montipora hodgsoni Veron, 2000	X	X
Montipora hoffmeisteri Wells, 1954	X	X
Montipora incrassata (Dana, 1846)	X	X
Montipora informis Bernard, 1897	X	X
Montipora mactanensis Nemenzo, 1979	X	X
Montipora malampaya Nemenzo, 1967	X	
Montipora meandrina (Ehrenberg, 1834)	X	X
Montipora millepora Crossland, 1952	X	X
Montipora mollis Bernard, 1897	X	X
Montipora monasteriata (Forskål, 1775)	X	X
Montipora niugini Veron, 2000	X	
Montipora nodosa (Dana, 1846)	X	X
Montipora orientalis Nemenzo, 1967	X	X
Montipora palawanensis Veron, 2000	X	X
Montipora peltiformis Bernard, 1897	X	X
Montipora porites Veron, 2000	X	X
Montipora samarensis Nemenzo, 1967	X	X
Montipora setosa Nemenzo, 1976	X	
Montipora spongodes Bernard, 1897	X	
Montipora spumosa (Lamarck, 1816)	X	
Montipora stellata Bernard, 1897	X	X
Montipora tuberculosa (Lamarck, 1816)	X	X
Montipora turgescens Bernard, 1897	X	X
Montipora turtlensis Veron and Wallace, 1984	X	
Montipora undata Bernard, 1897		X
Montipora venosa (Ehrenberg, 1834)	X	
Montipora verrucosa (Lamarck, 1816)	X	X
Montipora verruculosus Veron, 2000	X	X
Montipora vietnamensis Veron, 2000		X
Anacropora forbesi Ridley, 1884	X	X

Zooxanthellate Scleractinia	Recorded in eastern Indonesia (Veron, 2000)[1]	Recorded in Raja Ampat Islands
Anacropora matthai Pillai, 1973	X	X
Anacropora pillai Veron, 2000	X	
Anacropora puertogalerae Nemenzo, 1964	X	X
Anacropora reticulata Veron and Wallace, 1984	X	X
Anacropora spinosa Rehberg, 1892	X	
Acropora abrolhosensis Veron, 1985	X	X
Acropora abrotanoides (Lamarck, 1816)	X	X
Acropora aculeus (Dana, 1846)	X	X
Acropora acuminata (Verrill, 1864)	X	X
Acropora akajimensis Veron, 1990		X
Acropora anthocercis (Brook, 1893)	X	X
Acropora aspera (Dana, 1846)	X	X
Acropora austera (Dana, 1846)	X	X
Acropora awi Wallace and Wolstenholme, 1998	X	X
Acropora batunai Wallace, 1997	X	X
Acropora bifurcata Nemenzo, 1971		X
Acropora brueggemanni (Brook, 1893)	X	X
Acropora carduus (Dana, 1846)	X	X
Acropora caroliniana Nemenzo, 1976	X	X
Acropora cerealis (Dana, 1846)	X	X
Acropora chesterfieldensis Veron and Wallace, 1984	X	
Acropora clathrata (Brook, 1891)	X	X
Acropora convexa (Dana, 1846)	X	X
Acropora cophodactyla (Brook, 1892)	X	
Acropora copiosa Nemenzo, 1967	X	X
Acropora crateriformis (Gardiner, 1898)	X	
Acropora cuneata (Dana, 1846)	X	X
Acropora cylindrica Veron and Fenner, 2000	X	
Acropora cytherea (Dana, 1846)	X	X
Acropora dendrum (Bassett-Smith, 1890)	X	X
Acropora derawanensis Wallace (1997)	X	X
Acropora desalwii Wallace, 1994	X	X
Acorpora digitifera (Dana, 1846)	X	X
Acropora divaricata (Dana, 1846)	X	X
Acropora donei Veron and Wallace, 1984	X	X
Acropora echinata (Dana, 1846)	X	X
Acropora efflorescens (Dana, 1846)	X	
Acropora elegans (Milne Edwards and Haime, 1860)		X
Acropora elseyi (Brook, 1892)	X	X
Acropora exquisite Nemenzo, 1971	X	X
Acropora fastigata Nemenzo, 1967	X	
Acropora fenneri Veron, 2000	X	
Acropora florida (Dana, 1846)	X	X
Acropora formosa (Dana, 1846)	X	X

Zooxanthellate Scleractinia	Recorded in eastern Indonesia (Veron, 2000)[1]	Recorded in Raja Ampat Islands
Acropora gemmifera (Brook, 1892)	X	X
Acropora globiceps (Dana, 1846)	X	X
Acropora gomezi Veron, 2000	X	
Acropora grandis (Brook, 1892)	X	X
Acropora granulosa (Milne Edwards and Haime, 1860)	X	X
Acropora hoeksemai Wallace, 1997	X	X
Acropora horrida (Dana, 1846)	X	X
Acropora humilis (Dana, 1846)	X	X
Acropora hyacinthus (Dana, 1846)	X	X
Acropora indonesia Wallace, 1997	X	X
Acropora inermis (Brook, 1891)	X	X
Acropora insignis Nemenzo, 1967	X	X
Acropora irregularis (Brook, 1892)		X
Acropora jacquelineae Wallace, 1994		X
Acropora kimbeensis Wallace, 1999	X	X
Acropora kirstyae Veron and Wallace, 1984	X	
Acropora latistella (Brook, 1891)	X	X
Acropora lianae Nemenzo, 1967	X	
Acropora listeri (Brook, 1893)	X	
Acropora loisetteae Wallace, 1994		X
Acropora lokani Wallace, 1994	X	X
Acropora longicyathus (Milne Edwards and Haime, 1860)	X	X
Acropora loripes (Brook, 1892)	X	X
Acropora lovelli Veron and Wallace, 1984	X	
Acropora lutkeni Crossland, 1952	X	
Acropora macrostoma (Brook, 1891)		X
Acropora meridiana Nemenzo, 1971	X	X
Acropora microclados (Ehrenberg, 1834)	X	X
Acropora microphthalma (Verrill, 1859)	X	X
Acropora millepora (Ehrenberg, 1834)	X	X
Acropora mirabilis (Quelch, 1886)	X	X
Acropora monticulosa (Brüggemann, 1879)	X	X
Acropora mulitacuta Nemenzo, 1967	X	
Acropora nana (Studer, 1878)	X	X
Acropora nasuta (Dana, 1846)	X	X
Acropora navini Veron, 2000	X	
Acropora nobilis (Dana, 1846)	X	X
Acropora orbicularis Brook, 1892	X	
Acropora palifera (Lamarck, 1816)	X	X
Acropora palmerae Wells, 1954	X	
Acropora paniculata Verrill, 1902	X	X
Acropora papillarae Latypov, 1992		X
Acropora parahemprichii Veron, 2000		X
Acropora parilis (Quelch, 1886)	X	X

Zooxanthellate Scleractinia	Recorded in eastern Indonesia (Veron, 2000)[1]	Recorded in Raja Ampat Islands
Acropora pectinatus Veron, 2000		X
Acropora pichoni Wallace, 1999	X	
Acropora pinguis Wells, 1950		X
Acropora plana Nemenzo, 1967		X
Acropora plumosa Wallace and Wolstenholme, 1998	X	X
Acropora polystoma (Brook, 1891)	X	X
Acropora prostrata (Dana, 1846)	X	X
Acropora proximalis Veron, 2000		X
Acropora pulchra (Brook, 1891)	X	X
Acropora rambleri (Bassett-Smith, 1890)	X	X
Acropora retusa (Dana, 1846)	X	
Acropora robusta (Dana, 1846)	X	X
Acropora rosaria (Dana, 1846)	X	X
Acropora russelli Wallace, 1994		X
Acropora samoensis (Brook, 1891)	X	X
Acropora sarmentosa (Brook, 1892)	X	X
Acropora scherzeriana (Brüggemann, 1877)		X
Acropora secale (Studer, 1878)	X	X
Acropora sekiseinsis Veron, 1990	X	
Acropora selago (Studer, 1878)	X	X
Acropora seriata (Ehrenberg, 1834)	X	
Acropora solitaryensis Veron and Wallace, 1984	X	X
Acropora speciosa (Quelch, 1886)	X	X
Acropora spicifera (Dana, 1846)	X	
Acropora stoddarti Pillai and Scheer, 1976	X	
Acropora striata (Verrill, 1866)		X
Acropora subglabra (Brook, 1891)	X	X
Acropora subulata (Dana, 1846)	X	X
Acropora tenella (Brook, 1892)	X	
Acropora tenuis (Dana, 1846)	X	X
Acropora teres (Verrill, 1866)	X	
Acropora tizardi (Brook, 1892)	X	
Acropora torihalimeda Wallace, 1994	X	
Acropora tortuosa (Dana, 1846)	X	
Acropora tumida (Verrill, 1866)	X	
Acropora turaki Wallace, 1994	X	
Acropora tutuilensis Hoffmeister, 1925	X	
Acropora valenciennesi (Milne Edwards and Haime, 1860)	X	X
Acropora valida (Dana, 1846)	X	X
Acropora variabilis (Klunzinger, 1879)		X
Acropora vaughani Wells, 1954	X	X
Acropora vermiculata Nemenzo, 1967	X	X
Acropora verweyi Veron and Wallace, 1984	X	X
Acropora walindii Wallace, 1999	X	X

Zooxanthellate Scleractinia	Recorded in eastern Indonesia (Veron, 2000)[1]	Recorded in Raja Ampat Islands
Acropora wallaceae Veron, 1990	X	
Acropora willisae Veron and Wallace, 1984	X	X
Acropora yongei Veron and Wallace, 1984	X	X
Astreopora cucullata Lamberts, 1980	X	
Astreopora expansa Brüggemann, 1877	X	X
Astreopora gracilis Bernard, 1896	X	X
Astreopora incrustans Bernard, 1896	X	
Astreopora listeri Bernard, 1896	X	X
Astreopora macrostoma Veron and Wallace, 1984	X	
Astreopora myriophthalma (Lamarck, 1816)	X	X
Astreopora ocellata Bernard, 1896	X	X
Astreopora randalli Lamberts, 1980	X	X
Astreopora suggesta Wells, 1954	X	X
Family Euphilliidae		
Catalaphyllia jardinei (Saville-Kent, 1893)	X	
Nemenzophyllia turbida Hodgson and Ross, 1981	X	
Euphyllia ancora Veron and Pichon, 1980	X	X
Euphyllia cristata Chevalier, 1971	X	X
Euphyllia divisa Veron and Pichon, 1980	X	X
Euphyllia glabrescens (Chamisso and Eysenhardt, 1821)	X	X
Euphyllia paraancora Veron, 1990	X	
Euphyllia paradivisa Veron, 1990		X
Euphyllia yaeyamaenisis (Shirai, 1980)	X	X
Physogyra lichentensteini (Milne Edwards and Haime, 1851)	X	X
Plerogyra discus Veron and Fenner, 2000	X	
Plerogyra simplex Rehberg, 1892	X	
Plerogyra sinuosa (Dana, 1846)	X	X
Family Oculinidae		
Galaxea acrehelia Veron, 2000	X	X
Galaxea astreata (Lamarck, 1816)	X	
Galaxea cryptoramosa Fenner and Veron, 2000		X
Galaxea fascicularis (Linnaeus, 1767)	X	X
Galaxea horrescens (Dana, 1846)	X	X
Galaxea longisepta Fenner and Veron, 2000	X	X
Galaxea pauciseptia Claerebaudt, 1990	X	X
Family Siderasteridae		
Coscinaraea columna (Dana, 1846)	X	X
Coscinaraea crassa Veron and Pichon, 1980	X	
Coscinaraea exesa (Dana, 1846)	X	X
Coscinaraea wellsi Veron and Pichon, 1980	X	
Psammocora contigua (Esper, 1797)	X	X
Psammocora digitata Milne Edwards and Haime, 1851	X	X
Psammocora explanulata Horst, 1922	X	X

Zooxanthellate Scleractinia	Recorded in eastern Indonesia (Veron, 2000)[1]	Recorded in Raja Ampat Islands
Psammocora haimeana Milne Edwards and Haime, 1851	X	X
Psammocora nierstraszi Horst, 1921	X	X
Psammocora obtusangula (Lamarck, 1816)	X	X
Psammocora profundacella Gardiner, 1898	X	X
Psammocora stellata Verrill, 1864		X
Psammocora superficialis Gardiner, 1898	X	X
Pseudosiderastrea tayami Yabe and Sugiyama, 1935	X	
Siderastrea savignyana Milne Edwards and Haime, 1850	X	
Family Agariciidae		
Coeloseris mayeri Vaughan, 1918	X	X
Gardineroseris planulata Dana, 1846	X	X
Leptoseris explanata Yabe and Sugiyama, 1941	X	X
Leptoseris foliosa Dineson, 1980	X	X
Leptoseris gardineri Horst, 1921	X	X
Leptoseris hawaiiensis Vaughan, 1907	X	X
Leptoseris incrustans (Quelch, 1886)	X	
Leptoseris mycetoseroides Wells, 1954	X	X
Leptoseris papyracea (Dana, 1846)	X	X
Leptoseris scabra Vaughan, 1907	X	X
Leptoseris solida (Quelch, 1886)	X	
Leptoseris striata Fenner and Veron, 2000	X	X
Leptoseris tubulifera Vaughan, 1907	X	X
Leptoseris yabei (Pillai and Scheer, 1976)	X	X
Pachyseris foliosa Veron, 1990	X	X
Pachyseris gemmae Nemenzo, 1955	X	X
Pachyseris involuta (Studer, 1877)	X	
Pachyseris rugosa (Lamarck, 1801)	X	X
Pachyseris speciosa (Dana, 1846)	X	X
Pavona bipartita Nemenzo, 1980	X	X
Pavona cactus (Forskål, 1775)	X	X
Pavona clavus (Dana, 1846)	X	X
Pavona danae Milne Edwards and Haime, 1860		X
Pavona decussata (Dana, 1846)	X	X
Pavona duerdeni Vaughan, 1907	X	X
Pavona explanulata (Lamarck, 1816)	X	X
Pavona frondifera (Lamarck, 1816)		X
Pavona maldivensis (Gardiner, 1905)	X	X
Pavona minuta Wells, 1954	X	X
Pavona varians Verrill, 1864	X	X
Pavona venosa (Ehrenberg, 1834)	X	X
Family Fungiidae		
Ctenactis albitentaculata Hoeksema, 1989	X	X
Ctenactis crassa (Dana, 1846)	X	X
Ctenactis echinata (Pallas, 1766)	X	X

Zooxanthellate Scleractinia	Recorded in eastern Indonesia (Veron, 2000)[1]	Recorded in Raja Ampat Islands
Cycloseris colini Veron, 2000		X
Cycloseris costulata (Ortmann, 1889)	X	X
Cycloseris curvata (Hoeksema, 1989)	X	X
Cycloseris cyclolites Lamarck, 1801	X	X
Cycloseris erosa (Döderlein, 1901)		X
Cycloseris hexagonalis (Milne Edwards and Haime, 1848)	X	X
Cycloseris patelliformis (Boschma, 1923)	X	X
Cycloseris sinensis (Milne Edwards and Haime, 1851)	X	X
Cycloseris somervillei (Gardiner, 1909)	X	X
Cycloseris tenuis (Dana, 1846)	X	X
Cycloseris vaughani (Boschma, 1923)	X	X
Diaseris distorta (Michelin, 1843)	X	
Diaseris fragilis Alcock, 1893	X	X
Cantharellus jebbi Hoeksems, 1993	X	
Fungia concinna Verrill, 1864	X	X
Fungia corona Döderlein, 1901	X	
Fungia danai Milne Edwards and Haime, 1851	X	X
Fungia fralinae Nemenzo, 1955	X	X
Fungia fungites (Linneaus, 1758)	X	X
Fungia granulosa Klunzinger, 1879	X	X
Fungia horrida Dana, 1846	X	X
Fungia klunzingeri Döderlein, 1901	X	X
Fungia moluccensis Horst, 1919	X	X
Fungia paumotensis Stutchbury, 1833	X	X
Fungia repanda Dana, 1846	X	X
Fungia scabra Döderlein, 1901	X	X
Fungia scruposa Klunzinger, 1879	X	X
Fungia scutaria Lamarck, 1801	X	X
Fungia spinifer Claereboudt and Hoeksema. 1987	X	X
Halomitra clavator Hoeksema, 1989		X
Halomitra meiere Veron and Maragos, 2000		X
Halomitra pileus (Linnaeus, 1758)	X	X
Heliofungia actiniformis (Quoy and Gaimard, 1833)	X	X
Herpolitha limax (Houttuyn, 1772)	X	X
Herpolitha weberi Horst, 1921	X	X
Lithophyllon mokai Hoeksema, 1989	X	X
Lithophyllon undulatum Rehberg, 1892	X	X
Podabacia crustacea (Pallas, 1766)	X	X
Podabacia motuporensis Veron, 1990	X	X
Polyphyllia talpina (Lamarck, 1801)	X	X
Sandalolitha dentata Quelch, 1884	X	X
Sandalolitha robusta Quelch, 1886	X	X
Zoopilus echinatus Dana, 1846	X	X

Zooxanthellate Scleractinia	Recorded in eastern Indonesia (Veron, 2000)[1]	Recorded in Raja Ampat Islands
Family Pectinidae		
Echinophyllia aspera (Ellis and Solander, 1788)	X	X
Echinophyllia costata Fenner and Veron, 2000	X	X
Echinophyllia echinata (Saville-Kent, 1871)	X	X
Echinophyllia echinoporoides Veron and Pichon, 1980	X	X
Echinophyllia orpheensis Veron and Pichon, 1980	X	X
Echinophyllia patula (Hodgson and Ross, 1981)	X	X
Echinophyllia pectinata Veron, 2000		X
Echinophyllia patula (Hodgson and Ross, 1982)	X	X
Echinomorpha nishihirai Veron, 1990		X
Mycedium elephatotus (Pallas, 1766)	X	X
Mycedium mancaoi Nemenzo, 1979	X	X
Mycedium robokaki Moll and Borel-Best, 1984	X	X
Oxypora crassispinosa Nemenzo, 1979	X	X
Oxypora glabra Nemenzo, 1959	X	X
Oxypora lacera Verrill, 1864	X	X
Pectinia alcicornis (Saville-Kent, 1871)	X	X
Pectinia aylini (Wells, 1935)		X
Pectinia elongata Rehberg, 1892	X	X
Pectinia lactuca (Pallas, 1766)	X	X
Pectinia maxima (Moll and Borel-Best, 1984)	X	X
Pectinia paeonia (Dana, 1846)	X	X
Pectinia pygmaeus Veron, 2000	X	
Pectinia teres Nemenzo and Montecillo, 1981	X	X
Family Merulinidae		
Paraclavarina triangularis (Veron and Pichon, 1980)	X	
Hydnophora bonsai Veron, 1990		X
Hydnophora exesa (Pallas, 1766)	X	X
Hydnophora grandis Gardiner, 1904	X	X
Hydnophora microconos (Lamarck, 1816)	X	X
Hydnophora pilosa Veron, 1985	X	X
Hydnophora rigida (Dana, 1846)	X	X
Merulina ampliata (Ellis and Solander, 1786)	X	X
Merulina scabricula Dana, 1846	X	X
Scapophyllia cylindrica Milne Edwards and Haime, 1848	X	X
Family Dendrophylliidae		
Turbinaria frondens (Dana, 1846)	X	X
Turbinaria irregularis, Bernard, 1896	X	X
Turbinaria mesenterina (Lamarck, 1816)	X	X
Turbinaria patula (Dana, 1846)	X	
Turbinaria peltata (Esper, 1794)	X	X
Turbinaria reniformis Bernard, 1896	X	X
Turbinaria stellulata (Lamarck, 1816)	X	X
Heteropsammia cochlea (Spengler, 1781)	X	

Zooxanthellate Scleractinia	Recorded in eastern Indonesia (Veron, 2000)[1]	Recorded in Raja Ampat Islands
Heterocyathus aequicostatus Milne Edwards and Haime, 1848	X	
Family Mussidae		
Micromussa amakusensis (Veron, 1990)	X	X
Micromussa minuta (Moll and Borel-Best, 1984)	X	X
Acanthastrea bowerbanki Milne Edwards and Haime, 1851	X	X
Acanthastrea brevis Milne Edwards and Haime, 1849	X	
Acanthastrea echinata (Dana, 1846)	X	X
Acanthastrea faviaformis Veron, 2000		X
Acanthastrea hemprichii (Ehrenberg, 1834)		X
Acanthastrea hillae Wells, 1955	X	
Acanthastrea ishigakiensis Veron, 1990	X	X
Acanthastrea lordhowensis Veron and Pichon, 1982		X
Acanthastrea regularis Veron, 2000	X	X
Acanthastrea rotundoflora Chevalier, 1975	X	X
Acanthastrea subechinata Veron, 2000		X
Australomussa rowleyensis Veron, 1985	X	X
Blastomussa merleti Wells, 1961	X	
Blastomussa wellsi Wijsman-Best, 1973	X	X
Cynarina lacrymalis (Milne Edwards and Haime, 1848)	X	X
Lobophyllia corymbosa (Forskål, 1775)	X	X
Lobophyllia dentatus Veron, 2000	X	X
Lobophyllia diminuta Veron, 1985	X	X
Lobophyllia flabelliformis Veron, 2000	X	X
Lobophyllia hataii Yabe and Sugiyama, 1936	X	X
Lobophyllia hemprichii (Ehrenberg, 1834)	X	X
Lobophyllia pachysepta Chevalier, 1975	X	
Lobophyllia robusta Yabe and Sugiyama, 1936	X	X
Lobophyllia serratus Veron, 2000		X
Scolymia australis (Milne Edwards and Haime, 1849)	X	
Scolymia vitiensis Brüggemann, 1877	X	
Symphyllia agaricia Milne Edwards and Haime, 1849	X	X
Symphyllia hassi Pillai and Scheer, 1976	X	X
Symphyllia radians Milne Edwards and Haime, 1849	X	X
Symphyllia recta (Dana, 1846)	X	X
Symphyllia valenciennesii Milne Edwards and Haime, 1849	X	X
Family Faviidae		
Caulastrea curvata Wijsman-Best, 1972	X	
Caulastrea echinulata (Milne Edwards and Haime, 1849)	X	
Caulastrea furcata Dana, 1846	X	X
Caulastrea tumida Matthai, 1928	X	
Cyphastrea agassizi (Vaughan, 1907)	X	X
Cyphastrea chalcidium (Forskål, 1775)	X	X
Cyphastrea decadia Moll and Borel-Best, 1984	X	X

Zooxanthellate Scleractinia	Recorded in eastern Indonesia (Veron, 2000)[1]	Recorded in Raja Ampat Islands
Cyphastrea japonica Yabe and Sugiyama, 1932	X	X
Cyphastrea microphthalma (Lamarck, 1816)	X	X
Cyphastrea ocellina (Dana, 1864)	X	X
Cyphastrea serailea (Forskål, 1775)	X	X
Diploastrea heliopora (Lamarck, 1816)	X	X
Echinopora ashmorensis Veron, 1990	X	
Echinopora gemmacea Lamarck, 1816	X	X
Echinopora hirsuitissima Milne Edwards and Haime, 1849	X	X
Echinopora horrida Dana, 1846	X	X
Echinopora lamellosa (Esper, 1795)	X	X
Echinopora mammiformis (Nemenzo, 1959)	X	X
Echinopora pacificus Veron, 1990	X	X
Favia danae Verrill, 1872	X	X
Favia favus (Forskål, 1775)	X	X
Favia helianthoides Wells, 1954	X	X
Favia laxa (Klunzinger, 1879)	X	X
Favia lizardensis Veron and Pichon, 1977	X	X
Favia maritima (Nemenzo, 1971)	X	X
Favia matthai Vaughan, 1918	X	X
Favia maxima Veron and Pichon, 1977	X	X
Favia pallida (Dana, 1846)	X	X
Favia rotumana (Gardiner, 1899)	X	X
Favia rotundata (Veron and Pichon, 1977)	X	X
Favia speciosa Dana, 1846	X	X
Favia stelligera (Dana, 1846)	X	X
Favia truncatus Veron, 2000	X	X
Favia veroni Moll and Borel-Best, 1984	X	X
Favia vietnamensis Veron, 2000	X	
Barabattoia laddi (Wells, 1954)	X	X
Favites abdita (Ellis and Solander, 1786)	X	X
Favites bestae Veron, 2000	X	X
Favites chinensis (Verrill, 1866)	X	X
Favites complanata (Ehrenberg, 1834)	X	X
Favites flexuosa (Dana, 1846)	X	X
Favites halicora (Ehrenberg, 1834)	X	X
Fabites halicora (Ehrenberg, 1834)	X	X
Favites micropentagona Veron, 2000	X	X
Favites paraflexuosa Veron, 2000		X
Favites pentagona (Esper, 1794)	X	X
Favites russelli (Wells, 1954)	X	X
Favites stylifera (Yabe and Sugiyama, 1937)	X	X
Favites vasta (Klunzinger, 1879)	X	X
Goniastrea aspera Verrill, 1905	X	X
Goniastrea australensis (Milne Edwards and Haime, 1857)	X	X

Zooxanthellate Scleractinia	Recorded in eastern Indonesia (Veron, 2000)[1]	Recorded in Raja Ampat Islands
Goniastrea edwardsi Chevalier, 1971	X	X
Goniastrea favulus (Dana, 1846)	X	X
Goniastrea minuta Veron, 2000	X	
Goniastrea palauensis (Yabe and Sugiyama, 1936)	X	
Goniastrea pectinata (Ehrenberg, 1834)	X	X
Goniastrea ramosa Veron, 2000		X
Goniastrea retiformis (Lamarck, 1816)	X	X
Leptastrea bewickensis Veron and Pichon, 1977	X	
Leptastrea bottae (Milne Edwards and Haime, 1849)		X
Leptastrea inaequalis Klunzinger, 1879	X	
Leptastrea pruinosa Crossland, 1952	X	X
Leptastrea purpurea (Dana, 1846)	X	X
Leptastrea transversa Klunzinger, 1879	X	X
Leptoria irregularis Veron, 1990	X	
Leptoria phrygia (Ellis and Solander, 1786)	X	X
Montastrea annuligera (Milne Edwards and Haime, 1849)	X	X
Montastrea colemani Veron, 2000	X	X
Montastrea curta (Dana, 1846)	X	X
Montastrea magnistellata Chevalier, 1971	X	X
Montastrea multipunctata Hodgson, 1985	X	
Montastrea salebrosa (Nemenzo, 1959)	X	X
Montastrea valenciennesi (Milne Edwards and Haime, 1848)	X	X
Moseleya latistellata Quelch, 1884	X	
Oulastrea crispata (Lamarck, 1816)	X	X
Oulophyllia bennettae (Veron, Pichon, 1977)	X	X
Oulophyllia crispa (Lamarck, 1816)	X	X
Oulophyllia levis (Nemenzo, 1959)		X
Platygyra acuta Veron, 2000		X
Platygyra contorta Veron, 1990	X	
Platygyra daedalea (Ellis and Solander, 1786)	X	X
Platygyra sp "green"		
Platygyra lamellina (Ehrenberg, 1834)	X	X
Platygyra pini Chevalier, 1975	X	X
Platygyra ryukyuensis Yabe and Sugiyama, 1936	X	X
Platygyra sinensis (Milne Edwards and Haime, 1849)	X	X
Platygyra verweyi Wijsman-Best, 1976	X	X
Plesiastrea versipora (Lamarck, 1816)	X	X
Family Trachyphyllidae		
Trachyphyllia geoffroyi (Audouin, 1826)	X	X
Family Poritidae		
Alveopora allingi Hoffmeister, 1925	X	
Alveopora catalai Wells, 1968	X	X
Alveopora daedalea (Forskål, 1775)	X	

Zooxanthellate Scleractinia	Recorded in eastern Indonesia (Veron, 2000)[1]	Recorded in Raja Ampat Islands
Alveopora fenestrata (Lamarck, 1816)	X	
Alveopora gigas Veron, 1985	X	X
Alveopora marionensis Veron and Pichon, 1982	X	X
Alveopora spongiosa Dana, 1846	X	X
Alveopora tizardi Bassett-Smith, 1890	X	
Alveopora verrilliana Dana, 1872	X	
Goniopora albiconus Veron, 2000		X
Goniopora burgosi Nemenzo, 1955		X
Goniopora columna Dana, 1846	X	X
Goniopora djiboutiensis Vaughan, 1907	X	X
Goniopora eclipsensis Veron and Pichon, 1982		X
Goniopora fruticosa Saville-Kent, 1893	X	X
Goniopora lobata Milne Edwards and Haime, 1860	X	X
Goniopora minor Crossland, 1952	X	X
Goniopora palmensis Veron and Pichon, 1982	X	X
Goniopora pandoraensis Veron and Pichon, 1982	X	X
Goniopora pendulus Veron, 1985		X
Goniopora polyformis Zou, 1980		X
Goniopora somaliensis Vaughan, 1907	X	X
Goniopora stokesi Milne Edwards and Haime, 1851	X	X
Goniopora stutchburyi Wells, 1955	X	X
Goniopora tenella (Quelch, 1886)	X	X
Goniopora tenuidens (Quelch, 1886)	X	X
Porites annae Crossland, 1952	X	X
Porites attenuata Nemenzo 1955	X	X
Porites australiensis Vaughan, 1918	X	X
Porites cumulatus Nemenzo, 1955	X	
Porites cylindrica Dana, 1846	X	X
Porites deformis Nemenzo, 1955	X	X
Porites densa Vaughan, 1918		X
Porites eridani, Umbgrove, 1940	X	
Porites evermanni Vaughan, 1907	X	X
Porites flavus Veron, 2000		X
Porites horizontalata Hoffmeister, 1925	X	X
Porites latistellata Quelch, 1886	X	X
Porites lichen Dana, 1846	X	X
Porites lobata Dana, 1846	X	X
Porites lutea Milne Edwards and Haime, 1851	X	X
Porites mayeri Vaughan, 1918	X	X
Porites monticulosa Dana, 1846	X	X
Porites murrayensis Vaughan, 1918	X	X
Porites napopora Veron, 2000		X
Porites negrosensis Veron, 1990		X
Porites nigrescens Dana, 1846	X	X

Zooxanthellate Scleractinia	Recorded in eastern Indonesia (Veron, 2000)[1]	Recorded in Raja Ampat Islands
Porites ornata Nemenzo, 1971	X	
Porites profundus Rehberg, 1892		X
Porites rugosa Fenner & Veron, 2000		X
Porites rus (Forskal, 1775)	X	X
Porites sillimaniana Nemenzo, 1976	X	X
Porites solida (Forskål, 1775)	X	X
Porites stephensoni Crossland, 1952	X	X
Porites tuberculosa Veron, 2000	X	X
Porites vaughani Crossland, 1952	X	X
Nine unidentified species		9
TOTAL 565	490	465

Appendix 2

Coral species recorded at individual sites in the Raja Ampat Islands

D. Fenner

SPECIES	SITE RECORDS
Family Astrocoeniidae	
Stylocoeniella armata (Ehrenberg, 1834)	1,6, 16, 37, 39, 42, 43
Stylocoeniella guentheri Bassett-Smith, 1890	1, 2b, 7, 8, 16, 20, 21, 23, 24, 42, 43, 44
Family Pocilloporidae	
Palauastrea ramosa Yabe & Sugiyama, 1941	11, 29, 31, 35, 42
Pocillopora ankeli Scheer & Pillai, 1974	7, 14
Pocillopora damicornis (Linnaeus, 1758)	1, 2a, 2b, 3, 4, 5, 6, 7, 8, 10, 13, 14, 15, 16, 18, 19, 20, 21, 22, 24, 27, 28, 29, 30, 31, 33, 34, 37, 38, 39, 40, 41, 42, 43, 44
Pocillopora eydouxi Milne Edwards & Haime, 1860	1, 2, 3, 7, 10, 13, 14, 16, 18, 20, 21, 22, 24, 25, 26, 28, 30, 31, 32, 33, 34, 36, 39, 41, 43, 44
Pocillopora meandrina Dana, 1846	1, 2a, 2b, 3, 6, 7, 10, 11, 13, 14, 15, 17, 18, 21, 22, 25, 28, 30, 31, 32, 36, 37, 39, 41
Pocillopora verrucosa (Ellis & Solander, 1786)	1, 2, 3, 5, 6, 7, 10, 11, 12, 13, 14, 17, 18, 20, 21, 22, 24, 25, 28, 30, 31, 32, 33, 34, 38, 39, 41, 42, 43, 44
Seriatopora aculeata Qluelch, 1886	10, 15, 31, 34, 37, 42
Seriatopora caliendrum Ehrenberg, 1834	1, 2, 4, 6, 10, 11, 12, 13, 15, 17, 18, 22, 23, 24, 25, 26, 27, 31, 33, 34, 37, 39, 41, 42, 43, 44
Seriatopora hystrix Dana, 1846	1, 2b, 4, 5, 6, 7, 10, 13, 15, 16, 1, 19, 22, 23, 34, 35, 36, 37, 39, 30, 31, 34, 35, 37, 38, 39, 40, 41, 42, 44
Stylophora subseriata Ehrenberg, 1834	1, 3, 6, 7, 11, 12, 14, 15, 16, 17, 30, 22, 23, 24, 25, 29, 30, 31, 33, 34, 35, 36, 37, 38, 39, 40, 41, 42, 43, 44
Family Acroporidae	
Acropora abrolhosensis Veron,1985	2b, 20, 24, 26, 31, 35, 39, 40, 42, 44
Acropora abrotanoides (Lamarck, 1816)	2a, 3, 7, 10, 14, 17, 21, 22, 25, 30, 32, 37, 39, 41
Acropora aculeus (Dana, 1846)	41, 42, 43
Acropora austera (Dana, 1846)	10, 15, 17, 18, 21, 30, 31, 32, 33, 37, 43
Acropora brueggemanni (Brook, 1893)	6, 11, 12, 13, 15, 20, 23, 24, 26, 29, 30, 31, 39, 41, 42, 44
Acropora carduus (Dana, 1846)	2b, 12, 13, 20, 40, 44
Acropora carolineana Nemenzo, 1976	19, 24
Acropora cerealis (Dana, 1846)	10, 13, 15, 21, 22, 24, 25, 29, 30, 32, 33, 36, 37, 39, 40, 42, 43
Acropora clathrata (Brook, 1891)	1, 2a, 2b, 7, 10, 13, 14, 15, 17, 18, 20, 21, 22, 24, 25, 26, 30, 32, 33, 34, 37, 40
Acropora cophodactyla (Brook, 1842)	2a, 7, 18, 21, 31, 32, 33, 34, 36, 39, 44
Acropora cuneata (Dana, 1856)	2, 3
Acropora cylindrica Veron & Fenner, 2000	27, 38
Acropora cytherea (Dana, 1846)	1, 3, 6, 7, 11, 12, 13, 14, 15, 17, 18, 20, 21, 24, 28, 30, 33, 36, 42
Acropora sp. 1 "danai-like"	39, 43
Acropora desalwii Wallace, 1994	Several
Acorpora digitifera (Dana, 1846)	1, 2a, 3, 6, 7, 14, 15, 17, 20, 21, 22, 24, 25, 30, 31, 33, 37, 39, 41, 42, 44

SPECIES	SITE RECORDS
Acropora divaricata (Dana, 1846)	1, 6, 7, 10, 13, 14, 15, 17, 18, 20, 24, 25, 26, 28, 39, 40, 41, 42, 43
Acropora echinata (Dana, 1846)	2b, 26, 35
Acropora efflorescens	several
Acropora eleguns (Milne Edwards and Haime, 1860)	35, 40, 41, 42
Acropora florida (Dana, 1846)	1, 2a, 2b, 3, 5, 6, 7, 10, 11, 13, 14, 15, 16, 17, 18, 21, 22, 24, 25, 26, 28, 30, 32, 33, 34, 35, 39, 40, 41, 42, 43, 44
Acropora formosa (Dana, 1846)	1, 2b, 7, 10, 13, 15, 21, 24, 29, 30, 35, 39, 42
Acropora gemmifera (Brook, 1892)	2a, 3, 6, 10, 11, 14, 24, 25, 30, 31, 33, 34, 36, 39, 42
Acropora globiceps (Dana, 1846)(?)	several
Acropora grandis (Brook, 1892)	2b
Acropora granulosa (Milne Edwards & Haime, 1860)	2, 3, 4, 6, 7, 14, 20, 22, 28, 30, 42
Acropora hoeksemai Wallace, 1997	10, 17, 22, 26
Acropora horrida (Dana, 1846)	2b, 6, 11, 12, 15, 20, 24, 26, 35, 40, 41
Acropora humilis (Dana, 1846)	1, 3, 6, 7, 11, 14, 15, 17, 20, 21, 22, 24, 25, 30, 31, 33, 37, 39, 40, 41, 42, 43, 44
Acropora hyacinthus (Dana, 1846)	2b, 6, 7, 10, 11, 13, 14, 15, 16, 17, 18, 20, 21, 22, 24, 25, 26, 28, 30, 31, 32, 33, 36, 37, 39, 40, 41, 42, 43, 44
Acropora indonesia Wallace, 1997	6, 10, 11, 20, 21, 24, 25, 28, 30, 31, 37, 41, 42, 43
Acropora kirstyae Veron & Wallace, 1984	29, 35, 40
Acropora latistella (Brook, 1891)	2b, 10, 15, 17, 21, 22, 24, 25, 29, 30, 32, 33, 36, 37, 39, 41, 42, 43
Acropora lokani Wallace, 1994	27, 35
Acropora longicyathus (Milne Edwards & Haime, 1860)	1, 2b, 6, 29, 40, 41, 44
Acropora loripes (Brook, 1892)	1, 2a, 2b, 3, 4, 7, 8, 10, 11, 12, 13, 14, 15, 17, 18, 21, 24, 25, 30, 32, 33, 34, 36, 37, 39, 40, 41, 42, 43
Acropora lutkeni Crossland, 1952(?)	25, 32, others
Acropora microphthalma (Verrill, 1859)	18
Acropora millepora (Ehrenberg, 1834)	1, 2a, 2b, 6, 7, 10, 12, 15, 18, 19, 21, 24, 25, 26, 28, 29, 30, 32, 33, 34, 37, 39, 40, 41, 42, 43, 44
Acropora monticulosa (Bruggemann, 1879)	1, 2, 7, 14, 24, 30, 31, 32, 33, 36, 39, 41, 44
Acropora mulitacuta Nemenzo, 1967	38
Acropora nana (Studer, 1878)	13, 30, 31, 32, 33, 36
Acropora nasuta (Dana, 1846)	1, 6, 11, 12, 14, 20, 22, 24, 25, 29, 30, 31, 33, 39, 40, 41, 42, 44
Acropora nobilis (Dana, 1846)	7, 14, 15, 17, 20, 21, 22, 24, 26, 37,
Acropora palifera (Lamarck, 1816)	1, 2b, 3, 6, 10, 11, 14, 15, 18, 19, 20, 22, 23, 24, 25, 26, 27, 30, 31, 34, 35, 37, 38, 39, 40, 41, 42, 43, 44
Acropora paniculata Verrill, 1902	10, 11, 14, 18, 20, 21, 22, 24, 25, 26, 30, 33, 34, 36, 37, 39, 41, 42, 43, 44
Acropora papillarae Latypov, 1992	6, 13, 40, 44
Acropora pinguis Wells, 1950	1, 30, 31, 33, 39, 44
Acropora plumosa Wallace & Wolstenholme, 1998	6, 26, 29, 35, 39, 42, 43, 44
Acropora pulchra (Brook, 1891)	6, 13, 42
Acropora robusta (Dana, 1846)	6, 7, 14, 17, 20, 21, 22, 24, 25, 28, 31, 32, 33, 36, 37, 39, 43, 44
Acropora rosaria (Dana, 1846)	27

SPECIES	SITE RECORDS
Acropora samoensis Brook, 1891)	4, 6, 12, 18, 20, 21 ,24, 25, 27, 29, 35, 39, 40, 42
Acropora secale (Studer, 1878)	6, 7, 22, 24, 31, 44
Acropora selago (Studer, 1878)	7, 14, 16, 20, 24, 25, 29, 39, 41, 43, 44
Acropora sp. "selago-like"	3, 4
Acropora solitaryensis Veron & Wallace, 1984	28, 30, 31, 32, 33, 34, 36, 37, 39
Acropora speciosa (Quelch, 1886)	13
Acropora sp. "subulata-like"	7, 18, 20, 21, 29, 30, 34, 39, 41, 42
Acropora subglabra (Brook, 1891)	7, 12, 13, 22, 39, 42, 44
Acropora tenuis (Dana, 1846)	2b, 3, 6, 7, 10, 11, 12, 13, 14, 15, 16, 17, 18, 20, 21, 22, 24, 25, 26, 30, 31, 32, 34, 35, 36, 37, 39, 40, 42, 43, 44
Acropora valenciennesi (Milne Edwards & Haime, 1860)	1, 2, 3, 7, 10, 11, 12, 13, 14, 15, 16, 17, 18, 20, 21, 22, 24, 25, 26, 28, 30, 32, 34, 37, 41, 43
Acropora valida (Dana, 1846)	17, 18, 20, 22, 30, 31, 32, 33, 34, 36, 39, 41, 44
Acropora vaughani Wells, 1954	2, 6, 7, 9, 11, 13, 14, 16, 17, 18, 20, 26, 41, 43
Acropora walindii Wallace, 1999	11, 29
Acropora yongei Veron & Wallace, 1984	2b, 6, 7, 10, 13, 15, 16, 17, 18, 19, 20, 22, 24, 25, 26, 31, 39, 43
Acropora sp "yongei-like"	32, 33, 39
Anacropora forbesi Ridley, 1884	3, 12, 44
Anacropora matthai Pillai, 1973	2b, 11, 18, 29, 44
Anacropora puertogalerae Nemenzo, 1964	2b, 12, 26, 29, 40
Anacropora reticulata Veron & Wallace, 1984	2b, 26, 29, 38
Anacropora spinosa Rehberg, 1892	2b
Astreopora expansa Bruggemann, 1877	1
Astreopora gracilis Bernard, 1896	2, 7, 10, 22, 27, 35, 36, 39
Astreopora listeri Bernard, 1896	19, 23, 35
Astreopora myriophthalma (Lamarck, 1816)	1, 2, 3, 10, 12, 13, 20, 21, 23, 24, 25, 30, 33, 34, 36, 39, 40, 42, 43, 44
Astreopora randalli Lamberts, 1980	7, 12, 16, 23, 35, 38, 42
Astreopora suggesta Wells, 1954	8, 18, 36
Montipora aequituberculata Bernard, 1897	8, 15, 39, 43
Montipora altisepta Nemenzo, 1967	9, 11, 13, 23, 26, 40
Montipora sp. "brown ridge"	20, 24, 32, 33, 34, 39, 43
Montipora cactus Bernard, 1897	2b, 12, 15, 16, 40, 41, 42, 43, 44
Montipora caliculata (Dana, 1846)	27, 31, 33, 34, 37, 39, 40, 43
Montipora capitata Dana, 1846	2b, 6, 7, 10, 11, 19, 20, 23, 24, 27, 30, 31, 33, 39, 40, 43
Montipora capricornis Veron, 1985	11, 12, 13, several others
Montipora cebuensis Nemenzo, 1976	2, 7, 11
Montipora confusa Nemenzo, 1967	1, 2a, 10, 12, 13, 14, 15, 16, 17, 18, 20, 21, 22, 24, 25, 26, 28, 29, 30, 32, 33, 34, 7, 39, 40, 41, 42, 43, 44
Montipora corbettensis Veron & Wallace, 1984	15, 18, 22, 41, 43
Montipora danae (Milne Edwards and Haime, 1851)	37
Montipora delicatula Veron, 2000	11, 12, 18
Montipora florida Nemenzo, 1967	2b, 4, 11, 12, 19, 26, 27, 29, 35, 40, 42
Montipora foliosa (Pallas, 1766)	11, 18, 20, 32, 37, 39
Montipora foveolata (Dana, 1846)	30, 31, 32, 33, 34, 36
Montipora gaimardi Bernard, 1897	2b, 5
Montipora hispida Dana, 1846	1, 2b, 6, 10, 11, 13, 15, 18, 20, 22, 24, 26, 29, 31, 32, 39, 40, 42

SPECIES	SITE RECORDS
Montipora informis Bernard, 1897	7, 10, 11, 13, 17, 26, 27, 30, 31, 39, 42
Montipora mactanensis Nemenzo, 1979	27
Montipora mollis Bernard, 1897	2a, 2b, 5, 10, 11, 15, 17, 18, 33
Montipora palawanensis Veron, 2000	1, 11, 13, 14, 16, 17, 18, 22, 31, 33, 41
Montipora peltiformis Bernard, 1897	10
Montipora samarensis Nemenzo, 1967	2b, 11, 27, 31
Montipora stellata Bernard, 1897	2b, 8, 9, 29
Montipora tuberculosa Lamarck, 1816)	1, 6, 7, 10, 11, 12, 13, 15, 17, 19, 20, 21, 22, 24, 25, 32, 34, 37, 39, 41, 42, 43, 44
Montipora undata Bernard, 1897	7, 10, 12, 13, 14, 15, 16, 18, 24, 30, 32, 34, 43
Montipora venosa (Ehrenberg, 1834)	10, 41
Montipora verrucosa (Lamarck, 1816)	6, 7, 11, 12, 16, 17, 30, 36, 38, 43
Montipora verruculosus Veron, 2000	12, 26, 29, 35, 34, 40, 42
Montipora vietnamensis Veron, 2000	11

Family Poritidae

Alveopora allingi Hoffmeister, 1925	15
Alveopora catalai Wells, 1968	26, 40, 42
Goniopora pendulus Veron, 1985	15, 33
Goniopora tenuidens (Quelch, 1886)	16
Porites annae Crossland, 1952	25
Porites attenuata Nemenzo 1955	17, 37
Porites cylindrica Dana, 1846	1, 2a, 2b, 3, 7, 9, 10, 11, 12, 15, 16, 17, 18, 19, 20, 23, 25, 26, 27, 28, 29, 31, 32, 33, 34, 35, 37, 38, 39, 40, 41, 42, 43
Porites densa Vaughan, 1918	35, 39, 42, and others
Porites evermanni Vaughan, 1907	18, 22, 24, 31, 32, 33, 34, 39, 43
Porites horizontalata Hoffmeister, 1925	8, 12, 19, 23, 26, 27, 38, 44
Porites lichen Dana, 1846	22, 33
Porites monticulosa Dana, 1846	27, 31, 38, 40, 41
Porites murrayensis Vaughan, 1918	several
Porites rugosa Fenner & Veron, 2000	6
Porites rus (Forskal, 1775)	1, 2b, 3, 8, 10, 11, 16, 17, 19, 20, 24, 25, 29, 31, 32, 35, 36, 37, 39, 40, 41, 42
Porites stephensoni Crossland, 1952	several
Porites vaughani Crossland, 1952	6, 12, 39, 41, 42

Family Siderasteridae

Coscinaraea columna (Dana, 1846)	20, 23, 29, 31, 32, 35, 36, 37
Psammocora contigua (Esper, 1797)	3, 4, 6, 10, 11, 12, 13, 14, 23, 41, 43, 44
Psammocora digitata Milne Edwards & Haime, 1851	1, 3, 4, 5, 10, 13, 15, 17, 20, 24, 26, 29, 35, 40, 41, 43, 44
Psammocora explanulata van der Horst, 1922	2a
Psammocora haimeana Milne Edwards & Haime, 1851	4, 10, 43
Psammocora nierstraszi van der Horst, 1921	1, 2, 6, 10, 11, 15, 17, 18, 21, 22, 24, 25, 26, 29, 32, 36, 38, 39, 41, 44
Psammocora profundacella Gardiner, 1898	1, 13, 15, 22, 25, 26, 29, 34, 40, 41, 44
Psammocora superficialis Gardiner, 1898	2b, 4, 8, 12, 13, 41, 42

Family Agariciidae

Coeloseris mayeri Vaughan, 1918	1, 2b, 3, 6, 17, 20, 24, 25, 27, 29, 30, 32, 33, 35, 37, 38, 39, 43, 44

SPECIES	SITE RECORDS
Gardineroseris planulata Dana, 1846	1, 2a, 2b, 3, 6, 7, 10, 13, 14, 15, 16, 17, 18, 20, 21, 22, 26, 29, 32, 33, 39, 40, 411, 42, 43
Leptoseris explanata Yabe & Sugiyama, 1941	2a, 2b, 6, 7, 8, 9, 13, 14, 16, 18, 19, 20, 22, 23, 28, 29, 35, 36, 39, 41, 42, 44
Leptoseris foliosa Dineson, 1980	27, 35
Leptoseris gardineri Horst, 1921	11, 19, 27, 29, 35, 38, 40
Leptoseris hawaiiensis Vaughan, 1907	1, 2a, 6, 7, 9, 14, 18, 20, 22, 33, 38, 39, 41, 42, 43, 44
Leptoseris mycetoseroides Wells, 1954	1, 2a, 2b, 6, 8, 9, 12, 17, 20, 22, 30, 39, 41, 42
Leptoseris papyracea (Dana, 1846)	2b, 11, 13, 16, 42, 44
Leptoseris scabra Vaughan, 1907	26, 27, 29, 39, 44
Leptoseris yabei (Pillai & Scheer, 1976)	6, 11, 20, 40
Pachyseris foliosa Veron, 1990	2b, 9, 12, 26, 40, 42, 44
Pachyseris gemmae Nemenzo, 1955	2b, 27, 35, 37, 39, 41, 42, 44
Pachyseris rugosa (Lamarck, 1801)	3, 6, 13, 16, 17, 22, 23, 31, 35, 38, 40, 41, 42
Pachyseris speciosa (Dana, 1846)	1, 2b, 3, 4, 5, 6, 7, 8, 9, 10, 11, 12, 13, 14, 15, 16, 17, 18, 19, 20, 21, 22, 23, 25, 26, 27, 28, 29, 30, 31, 33, 35, 36, 37, 38, 39, 40, 41, 42, 43, 44
Pavona bipartita Nemenzo, 1980	1, 2a, 5, 6, 15, 17, 28, 33, 35, 39, 40, 41, 44
Pavona cactus (Forskal, 1775)	1, 2b, 4, 6, 7, 9, 11, 12, 13, 16, 17, 18, 19, 23, 26, 29, 35, 37, 38, 40, 41, 42, 44
Pavona clavus (Dana, 1846)	1, 2b, 3, 4, 11, 13, 17, 18, 19, 26, 35, 37, 41, 44
Pavona danae Milne Edwards & Haime, 1860?	2b
Pavona decussata (Dana, 1846)	1, 3, 4, 10, 11, 13, 19, 21, 25, 27, 29, 32, 42, 43, 44
Pavona duerdeni Vaughan, 1907	17
Pavona sp "duerdeni-like"	1, 5, 10, 14, 17, 21, 22, 24, 25, 32, 34, 37, 39, 40, 44
Pavona explanulata (Lamarck, 1816)	2, 5, 6, 7, 8, 10, 12, 13, 14, 15, 16, 17, 20, 22, 24, 27, 29, 30, 31, 32, 35, 37, 38, 39, 40, 41, 42, 43, 44
Pavona minuta Wells, 1954	13, 28, 29, 41
Pavona varians Verrill, 1864	1, 2b, 3, 6, 7, 8, 9, 10, 11, 12, 13, 14, 15, 16, 17, 18, 19, 20, 21, 22, 24, 25, 26, 29, 30, 31, 32, 33, 34, 35, 37, 38, 39, 40, 41, 42, 43, 44
Pavona venosa (Ehrenberg, 1834)	3, 4, 6, 8, 9, 19, 25, 27, 29, 31, 37, 41, 44
Family Fungiidae	
Ctenactis albitentaculata Hoeksema, 1989	2b, 6, 11, 12, 13, 16, 27, 29, 31, 35, 37, 40, 42
Ctenactis crassa (Dana, 1846)	2b, 6, 11, 12, 16, 27, 33, 40, 41, 42, 44
Ctenactis echinata (Pallas, 1766)	1, 4, 7, 10, 11, 13, 15, 16, 18, 20, 21, 22, 25, 26, 27, 29, 37, 40, 42, 43, 44
Cycloseris colini Veron, 2000	38
Cycloseris cyclolites Lamarck, 1801	8, 13
Cycloseris somervillei (Gardiner, 1909)	35
Cycloseris vaughani (Boschma, 1923)	1, 18, 27
Diaseris fragilis Alcock, 1893	29
Fungia concinna Verrill, 1864	2b, 3, 4, 8, 12, 15, 23, 39, 43
Fungia fralinae Nemenzo, 1955	27, 29, 40, 42
Fungia fungites (Linneaus, 1758)	1, 2b, 6, 7, 12, 13, 15, 16, 18, 20, 24, 25, 26, 29, 31, 37, 39, 40, 42, 43, 44
Fungia granulosa Klunzinger, 1879	2b, 10, 15, 18, 21, 22, 27, 33, 34, 37, 42, 43
Fungia horrida Dana, 1846	1, 2a, 2b, 4, 6, 7, 9, 11, 12, 13, 15, 22, 25, 29, 31, 37, 39, 40, 41, 42, 44
Fungia klunzingeri Doderlein, 1901	1, 2b, 4, 7, 11, 13, 16, 18, 27, 31, 37, 39
Fungia moluccensis Horst, 1919	19, 29, 42

 Rapid Assessment Program

SPECIES	SITE RECORDS
Fungia paumotensis Stutchbury, 1833	1, 2a, 2b, 3, 4, 6, 8, 9, 11, 12, 13, 15, 16, 18, 20, 22, 23, 24, 25, 26, 27, 29, 31, 33, 35, 37, 38, 40, 41, 42, 43
Fungia repanda Dana, 1846	2b, 7, 22, 24, 41
Fungia scruposa Klunzinger, 1816	2b, 4, 6, 7, 8, 9, 10, 12, 13, 16, 18, 20, 23, 25, 29, 31, 39, 40, 42, 43, 44
Fungia scutaria Lamarck, 1816	1, 6, 7, 10, 13, 14, 18, 20, 21, 25, 31, 32, 33, 34, 37, 39, 43
Halomitra clavator Hoeksema, 1989	11, 27
Halomitra meiere Veron & Maragos, 2000	34, 39
Halomitra pileus (Linnaeus, 1758)	1, 2b, 4, 6, 7, 11, 12, 13, 16, 17, 18, 25, 26, 30, 31, 40, 41, 43, 44
Heliofungia actiniformis Quoy & Gaimard, 1837	2b, 4, 8, 15, 16, 17, 22, 23, 26, 27, 29, 30, 31, 35, 40, 42, 44
Herpolitha limax (Houttuyn, 1772)	1, 2b, 3, 6, 7, 8, 9, 10, 11, 12, 13, 14, 16, 17, 18, 19, 20, 22, 23, 25, 27, 29, 30, 31, 33, 35, 39, 40, 41, 42, 43, 44
Herpolitha weberi Horst, 1921	27
Lithophyllon mokai Hoeksema, 1989	27, 35, 38, 40
Podabacia crustacea (Pallas, 1766)	3, 4, 6, 11, 12, 16, 19, 21, 25, 26, 35, 40, 42
Podabacia motuporensis Veron, 1990	1, 7, 8, 12, 16, 21, 22, 23, 29, 41, 44
Polyphyllia talpina Lamarck, 1801	2b, 3, 7, 11, 12, 21, 24, 29, 34, 40
Sandalolitha dentata Quelch, 1884	14
Sandalolitha robusta Quelch, 1886	1, 2a, 2b, 3, 4, 6, 7, 8, 10, 11, 12, 13, 14, 15, 16, 17, 18, 20, 21, 22, 24, 26, 27, 28, 31, 33, 34, 35, 36, 37, 39, 42, 44
Zoopilus echinatus Dana, 1846	11, 16, 26, 29, 31, 40, 42, 44
Family Oculinidae	
Galaxea astreata (Lamarck,. 1816)	4, 8, 11, 12, 13, 15, 16, 18, 22, 25, 26, 27, 29, 32, 33, 37, 40, 41, 42, 43, 44
Galaxea fascicularis (Linnacus, 1767)	1, 2, 3, 4, 6, 7, 8, 10, 11, 12, 13, 14, 15, 16, 17, 18, 20, 21, 22, 23, 24, 25, 26, 27, 28, 29, 30, 31, 32, 33, 34, 35
Galaxea horrescens (Dana, 1846)	2b, 9, 11, 13, 19,, 26, 27, 31, 38, 40, 41
Galaxea paucisepta Claerebaudt, 1990	1, 2b, 3, 6, 11, 12, 26, 27, 29, 35, 40
Family Pectinidae	
Echinophyllia aspera (Ellis & Solander, 1788)	1, 3, 6, 11, 12, 13, 17, 21, 23, 25, 31, 34, 35, 37, 39, 40, 42
Echinophyllia costata Fenner & Veron, 2000	11, 26, 27, 40, 42
Echinophyllia echinata (Saville-Kent, 1871)	27, 35
Echinophyllia echinoporoides Veron & Pichon, 1979	4, 9, 23
Echinophyllia orpheensis Veron & Pichon, 1980	23
Echinophyllia patula (Hodgson & Ross, 1982)	1, 6, 7, 11, 13, 16, 18, 20, 26, 28, 31, 37, 39, 40, 41, 42
Mycedium elephatotus (Pallas, 1766)	1, 2, 3, 6, 7, 8, 9, 12, 13, 15, 16, 17, 18, 20, 21, 22, 28, 31, 32, 34, 36, 39, 42, 44
Mycedium mancaoi Nemenzo, 1979	2, 6, 12, 31, 37, 39, 40, 41, 42
Mycedium robokaki Moll & Borel-Best, 1984	1, 2b, 3, 6, 11, 15, 16, 17, 18, 20, 21, 22, 24, 26, 27, 28, 29, 30, 31, 33, 35, 36, 37, 38, 39, 40, 41, 42, 44
Oxypora crassispinosa Nemenzo, 1979	1, 2b, 6, 7, 9, 10, 11, 12, 16, 18, 20, 31, 35, 37, 38, 40, 42, 44
Oxypora glabra Nemenzo, 1959	11, 19, 27, 29, 35, 40
Oxypora lacera Verrill, 1864	1, 2a, 2b, 6, 7, 11, 12, 15, 16, 18, 20, 21, 22, 23, 25, 26, 28, 30, 31, 34, 36, 39, 40, 41, 42, 44

SPECIES	SITE RECORDS
Pectinia aylini (Wells, 1935)	1, 12, 23, 25, 31, 42
Pectinia lactuca (Pallas, 1766)	1, 2a, 2b, 6, 7, 9, 10, 12, 13, 15, 16, 17, 18, 20, 21, 22, 23, 25, 28, 29, 30, 31, 33, 36, 39, 40, 41, 42, 44
Pectinia maxima (Moll and Borel-Best, 1984)	9, 23
Pectinia paeonia (Dana, 1846)	12, 19, 27, 42
Pectinia teres Nemenzo, 1981	2b, 4, 6, 8, 9, 11, 12, 23, 38, 40, 41, 42
Family Mussidae	
Acanthastrea brevis Milne Edwards and Haime, 1849	31, 35
Acanthastrea echinata (Dana, 1846)	24, 26, 30, 36
Acanthastrea hemprichii (Ehrenberg, 1834)	36
Acanthastrea subechinata Veron, 2000	29, 33, 34
Australomussa rowleyensis Veron, 1985	2b, 4, 6, 7, 12, 20, 23, 27, 29, 38, 42, 43
Blastomussa wellsi Wijsman-Best, 1973	29
Cynarina lacrymalis (Milne Edwards & Haime, 1848)	4, 29, 35
Lobophyllia corymbosa Forskal, 1775	27, 35, 42
Lobophyllia flabelliformis Veron, 2000	4, 7, 12, 14, 40
Lobophyllia hataii Yabe & Sugiyama, 1936	18
Lobophyllia hemprichii (Ehrenberg, 1834)	1, 2b, 3, 4, 6, 8, 9, 13, 14, 16, 17, 18, 19, 20, 23, 24, 25, 26, 27, 29, 30, 31, 32, 34, 35, 36, 37, 38, 39, 40, 42, 43
Lobophyllia robusta Yabe & Sugiyama, 1936	2b, 3, 5, 7, 10, 13, 16, 17, 18, 20, 21, 31, 32, 34, 39, 40, 41, 42
Scolymia vitiensis Haime, 1852	38
Symphyllia agaricia Milne Edwards & Haime, 1849	1, 2b, 3, 6, 7, 13, 14, 15, 16, 17, 18, 20, 21, 22, 24, 28, 31, 32, 33, 34, 35, 36, 37, 39, 41, 44
Symphyllia hassi Pillai & Scheer, 1976	12, 41, 42, 44
Symphyllia radians Milne Edwards & Haime, 1849	1, 3, 7, 8, 10, 13, 14, 15, 16, 17, 18, 20, 22, 24, 25, 28, 30, 31, 32, 33, 34, 36, 37, 39, 40, 41, 42, 44
Symphyllia recta (Dana, 1846)	2, 3, 7, 8, 11, 12, 14, 20, 21, 22, 23, 24, 25, 30, 32, 33, 37, 38, 39, 40, 43
Symphyllia valenciennesii Milne Edwards & Haime, 1849	2, 13
Family Merulinidae	
Hydnophora exesa (Pallas, 1766)	3, 4, 7, 11, 12, 13, 15, 16, 23, 27, 28, 29, 30, 31, 32, 33, 40, 42, 44
Hydnophora grandis Gardiner, 1904	3, 7, 11, 13, 15, 19, 20, 23, 25, 26, 29, 31, 36, 37, 40, 42, 43
Hydnophora microconos (Lamarck, 1816)	1, 2a, 3, 6, 7, 10, 111, 13, 14, 15, 16, 17, 18, 20, 21, 22, 24, 25, 28, 30, 31, 32, 33, 34, 36, 37, 39, 41, 43, 44
Hydnophora pilosa (Veron, 1985*)*	2b, 11, 15, 16, 18, 26, 28, 32, 33, 34, 36, 40
Hydnophora rigida (Dana, 1846)	1, 2a, 2b, 3, 4, 9, 10, 11, 12, 13, 15, 16, 17, 20, 21, 24, 2, 30, 35, 37, 38, 40, 41, 42, 43, 44
Merulina ampliata (Ellis & Solander, 1786)	1, 2, 3, 4, 6, 7, 8, 9, 11, 12, 13, 14, 19, 22, 23, 24, 25, 26, 27, 33, 35, 36, 38, 39, 40, 41, 42, 44
Merulina scabricula Dana, 1846	1, 3, 7, 10, 11, 12, 13, 15, 17, 18, 20, 21, 22, 23, 24, 29, 30, 311, 32, 34, 36, 37, 39, 40, 41, 43, 44
Scapophyllia cylindrica Milne Edwards & Haime, 1848	3, 22, 24, 25, 26, 32

SPECIES	SITE RECORDS
Family Faviidae	
Caulastrea echinulata (Milne Edwards & Haime, 1849)	16, 24, 40
Caulastrea furcata Dana, 1846	16, 35
Cyphastrea agassizi (Vaughan, 1907)	38
Cyphastrea decadia Moll and Borel-Best, 1984	2b, 11
Diploastrea heliopora (Lamarck, 1816)	2a, 2b, 4, 5, 6, 7, 8, 9, 10, 12, 13, 14, 15, 16, 17, 18, 19, 20, 21, 22, 24, 25, 26, 28, 29, 30, 31, 32, 33, 35, 36, 37, 38, 39, 40, 41, 42, 43
Echinopora gemmacea Lamarck, 1816	2b, 7, 10, 13, 15, 16, 18, 20, 21, 22, 24, 25, 26, 28, 31, 32, 34, 35, 37, 42
Echinopora hirsuitissima Milne Edwards & Haime, 1849	7, 10, 16, 17, 18, 20, 25, 28, 30, 31, 32, 33, 34, 35, 36, 39
Echinopora horrida Dana, 1846	2b, 11, 13, 25, 26, 31, 35, 39, 40, 41, 42
Echinopora lamellosa (Esper, 1795)	1, 2b, 3, 5, 6, 7, 9, 12, 15, 23, 24, 26, 27, 29, 37, 40, 41, 42, 44
Echinopora mammiformis (Nemenzo, 1959)	2b, 11, 12, 19, 23, 27, 35, 38, 39
Echinopora pacificus Veron, 1990	35, 42
Favia laxa (Klunzinger, 1879)	19, 35, 38
Favia maxima Veron & Pichon, 1977	1, 21
Favia pallida (Dana, 1846)	6, 7, 11, 15, 16, 20, 24, 25, 29, 32, 39, 44
Favia rotundata Veron & Pichon, 1977	4, 6, 12, 23, 29, 35, 38
Favia stelligera (Dana, 1846)	1, 2b, 3, 6, 7, 13, 17, 18, 20, 21, 25, 26, 31, 32, 35, 36, 37, 39, 40, 41, 43, 44
Favia truncatus Veron, 2000	11, 43
Favites abdita (Ellis & Solander, 1786)	1, 3, 6, 10, 11, 12, 14, 15, 16, 18, 19, 20, 21, 23, 24, 25, 30, 31, 32, 37, 38, 39, 40, 41, 42, 43, 44
Favites halicora (Ehrenberg, 1834)	1, 4, 12, 19, 23, 35, 43, 44
Favites paraflexuosa Veron, 2000	7, 15, 17, 20, 24, 25, 30, 31, 32, 33
Favites pentagona (Esper, 1794)	43
Goniastrea edwardsi Chevalier, 1971	3, 7, 12, 13, 19, 23, 25
Goniastrea pectinata (Ehrenberg, 1834)	1, 2b, 3, 6, 8, 10, 12, 15, 22, 23, 24, 26, 30, 31, 32, 33, 36, 37, 40, 44
Goniastrea ramosa Veron, 2000	44
Goniastrea retiformis (Lamarck, 1816)	2b, 8, 10, 11, 12, 15, 22, 23, 24, 26, 30, 31, 32, 33, 36, 37, 39, 40, 44
Leptastrea bewickensis Veron & Pichon, 1977	
Leptastrea pruinosa Crossland, 1952	1, 3, 14, 23, 25, 30, 35, 36, 42
Leptastrea purpurea (Dana, 1846)	2a, 2b, 4, 8, 12, 19, 23, 24, 27, 31, 38, 39, 43, 44
Leptastrea transversa Klunzinger, 1879	1, 4, 6, 7, 19, 20, 30, 39
Leptoria phrygia (Ellis & Solander)	3, 13, 14, 15, 16, 17, 20, 22, 24, 30, 32, 34, 36, 39, 40, 41, 43
Montastrea curta (Dana, 1846)	22, 31, 36, 44
Montastrea magnistellata Chevalier, 1971	3, 15, 18, 23, 24, 26, 43
Montastrea salebrosa (Nemenzo, 1959)	35, 42
Oulastrea crispata (Lamarck, 1816)	4, 13, 14
Oulophyllia bennettae Veron, Pichon, & Wijsman-Best, 1977	4, 12, 20, 27, 29, 33, 35, 40
Oulophyllia crispa (Lamarck, 1816)	3, 6, 11, 12, 13, 14, 15, 16, 17, 18, 20, 21, 24, 25, 26, 28, 29, 31, 32, 33, 34, 37, 40, 41, 42, 43, 44

SPECIES	SITE RECORDS
Platygyra acuta Veron, 2000	7, 10, 14, 15, 21, 32, 33
Platygyra daedalea (Ellis & Solander, 1786)	1, 2a, 2b, 3, 6, 7, 10, 12, 13, 17, 18, 20, 22, 24, 25, 26, 30, 31, 34, 36, 37, 39, 40, 41, 42, 43, 44
Platygyra sp "green"	16, 27
Platygyra lamellina (Ehrenberg, 1834)	11, 12, 17, 19, 23, 27, 35, 38
Platygyra sinensis (Milne Edwards & Haime, 1849)	15
Platygyra verweyi Wijsman-Best, 1976	23
Plesiastrea versipora (Lamarck, 1816)	1, 2, 8, 10, 15, 16, 17, 21, 28, 30, 32, 34, 37, 38, 39, 41

Family Trachyphyllidae

Trachyphyllia geoffroyi Audouin, 1826	4, 29

Family Euphilliidae

Euphyllia ancora Veron & Pichon, 1979	2b, 7, 12, 15, 16, 26, 29, 40, 41, 42
Euphyllia glabrescens (Chamisso & Eysenhardt, 1821)	4, 11, 15, 16, 18, 26, 28, 29, 40, 42, 43
Euphyllia paradivisa Veron, 1990	4, 15
Euphyllia yaeyamaenisis (Shirai, 1980)	16, 29, 38, 40, 42
Physogyra lichentensteini Milne Edwards & Haime, 1786	2, 4, 6, 7, 8, 11, 12, 14, 15, 16, 18, 19, 20, 21, 22, 25, 26, 27, 28, 29, 31, 39, 40, 41, 42, 43
Plerogyra sinuosa (Dana, 1846)	4, 6, 8, 12, 16, 23, 27, 29, 33, 35, 40, 41

Family Dendrophylliidae

Dendrophyllia coccinea	14, 18, 39
Rhizopsammia verrilli	1, 2, 10, 14, 18, 28, 32, 39, 44
Tubastraea coccinea Lesson, 1829	1, 2, 5, 10, 14, 18, 21, 25, 28, 32, 33, 37, 39, 43
Tubastraea micranthus Ehrenberg, 1834	1, 2, 5, 7, 10, 13, 14, 17, 18, 20, 21, 28, 33, 39, 42, 43
Turbinaria frondens Dana, 1846	2, 10, 13, 14, 15, 17, 18, 21, 28, 34, 39
Turbinaria mesenterina (Lamarck, 1816)	7, 12, 13, 14, 16, 17, 18, 20, 22, 24, 27, 29, 31, 35, 36, 40, 42, 44
Turbinaria peltata (Esper, 1794)	1, 5, 7, 10, 13, 14, 15, 17, 18, 20, 21, 22, 28, 29, 34, 36, 37, 43
Turbinaria reniformis Bernard, 1896	2, 6, 12, 16, 28, 35, 36, 42
Turbinaria stellulata (Lamarck, 1816)	7, 30, 33, 35, 38

Family Heliporidae

Heliopora coerulea	3, 5, 9, 10, 11, 15, 26, 29, 30, 31, 36, 37, 39, 42, 44

Family Clavulariidae

Tubipora musica Linnaeus, 1758	3, 13, 15, 17, 21, 29, 30
Tubipora sp. 1 "large feathery"	11, 13, 27, 39, 41, 42
Tubipora sp. 2 "grey center"	13, 33
Tubipora sp. 3 "smooth large"	13, 15, 16, 37

Family Milleporidae

Millepora dichotoma	2, 5, 9, 10, 13, 22, 30, 31, 35, 37
Millepora exaesa	2, 3, 6, 7, 9, 10, 13, 14, 19, 21, 22, 24, 25, 28, 29, 30, 31, 34, 35, 36, 37, 38, 39, 41, 43, 44
Millepora intricata	3, 6, 7, 10, 11, 12, 13, 15, 17, 22, 26, 28, 29, 30, 31, 32, 33, 37, 39, 40, 41, 42, 44
Millepora platyphylla	2, 6, 7, 10, 15, 17, 20, 25, 31, 33, 37, 39, 41, 43

Family Stylasteridae

Stylaster sp. 1 orange or pink	7, 10, 13, 14, 17, 18, 20, 22, 25, 30, 39, 43
Distichopora violacea (Ellis & Solander, 1788)	14, 28, 30, 32, 33, 34, 36

Appendix 3

Molluscs recorded at the Raja Ampat Islands.

The letter "B" appearing after the site number indicates that specimen was collected in beach drift

F.E. Wells

SPECIES	SITE RECORDS
CLASS POLYPLACOPHORA	
Family Chitonidae	
Acanthopleura gemmata (Blainville, 1825)	1, 4, 5, 16 ,20 ,23 ,24 ,28 ,31 ,38 ,42 ,44
Acanthopleura spinosa (Bruguière, 1792)	5
Tonica lamellosa (Quoy & Gaimard, 1835)	3
CLASS GASTROPODA	
Family Patellidae	
Cellana rota (Gmelin, 1791)	20, 21, 25, 44
Cellana testudinaria (Linnaeus, 1758)	38, 44
Patella flexuosa Quoy & Gaimard, 1834	19
Family Acmaeidae	
Patelloida conodialis (Pease, 1868)	1
Patelloida saccharina (Linnaeus, 1758)	6, 13, 19, 20, 44
Patelloida striata (Quoy & Gaimard, 1834)	20,25 ,32 ,38
Family Haliotidae	
Haliotis asinina Linnaeus, 1758	2a, 12, 17, 44
Haliotis crebrisculpta Sowerby, 1914	6, 40
Haliotis ovina Gmelin, 1791	1-3, 5, 11, 16-18, 20, 35, 37, 41, 42, 44
Haliotis planata Sowerby, 1833	2a, 7
Haliotis varia Linnaeus, 1758	2b, 11
Family Fissurellidae	
Diodora singaporensis (Reeve, 1850)	2b
Hemitoma panhi (Quoy and Gaimard, 1834)	27
Suctus unguis (Linnaeus, 1758)	13, 22, 39, 40
Family Turbinidae	
Astralium calcar (Linnaeus, 1758)	23, 26, 29, 34
Astralium rhodostomum (Lamarck, 1822)	11, 13, 16, 17, 39
Bolma erectospinosa (Habe & Okutani, 1980)	16B
Monodonta labio (Linnaeus, 1758)	8
Phasianella solida (Born, 1778)	25
Turbo argyrostomus (Linnaeus, 1758)	1, 2, 4, 5, 7, 11, 13, 14, 16, 17, 21-23, 25, 33, 34, 41, 43
Turbo chrysostoma (Linnaeus, 1758)	30
Turbo cinereus Born, 1778	16B
Turbo petholatus Linnaeus, 1758	1, 2b, 12, 15, 16, 24, 27, 29, 31, 32, 34-36, 39, 41, 43, 44
Turbo radiatus Gmelin, 1791	26, 30, 36, 40

SPECIES	SITE RECORDS
Family Trochidae	
Angaria delphinus (Linnaeus, 1758)	9, 19b ,29
Calthalotia sp.	13
Cantharidus sp.	37
Chrysostoma paradoum (Born, 1780)	2a 9, 10-12, 22, 25, 26, 29, 35, 44
Clanculus atropurpureus (Gould, 1849)	16B
Clanculus margaritius Philippi, 1847	42
Ethalia cf. *guamensis* (Quoy & Gaimard, 1834)	25, 27
Euchelus cf. *instriticus* (Gould, 1849)	42
Herpetopoma atrata (Gmelin, 1791)	38, 42
Liotina peronii (Kiener, 1839)	16B
Stomatella varia (A. Adams, 1850)	37, 40, 44
Stomatia phymotis Helbling, 1779	4b, 18, 44
Tectus conus (Gmelin, 1791)	5, 7, 10, 19, 23, 27, 38, 40
Tectus fenestratus Gmelin, 1790	2b, 4, 6, 16, 27, 38
Tectus maculatus Linnaeus, 1758	17, 21, 23, 32, 38
Tectus niloticus Linnaeus, 1767	2a, 7, 15, 21, 26, 38, 40
Tectus pyramis Born, 1778	1-3, 6, 7, 10-13, 15-19, 21, 22, 24, 26, 27, 33-35, 40-42, 44
Tectus triserialis (Lamarck, 1822)	38, 40
Trochus hanleyanus (Reeve, 1843)	27
Trochus lacianatus	40, 41
Trochus sp.	31
Family Neritidae	
Nerita albicilla Linnaeus, 1758	6, 16, 32, 44
Nerita chamaeleon Linnaeus, 1758	39
Nerita costata Gmelin, 1791	39, 42, 44
Nerita plicata Linnaeus, 1758	1, 6, 13, 15, 16, 26, 32, 37, 39, 44
Nerita polita Linnaeus, 1758	16, 39, 44
Nerita reticulata Karsten, 1789	16B, 44
Nerita undata Linnaeus, 1758	16, 18, 23, 38, 42, 44
Family Cerithiidae	
Cerithium alveolus Hombron & Jacquinot, 1854	13, 44
Cerithium balteatum Philippi, 1848	2b, 6, 7, 12, 13, 25, 27, 35, 38, 40-43
Cerithium columna Sowerby, 1834	1, 2a ,6, 10, 21, 39, 41
Cerithium echinatum (Lamarck, 1822)	44
Cerithium lifuense Melvill & Standen, 1895	27, 42, 44
Cerithium munitum Sowerby, 1855	9, 19, 42
Cerithium nesioticum Pilsbry & Vanetta, 1906	2, 16, 18, 38
Cerithium nodulosus (Bruguière, 1792)	2a, 2b, 6, 7, 11, 18, 20, 21, 26, 30, 33, 38
Cerithium rostratum Soweιby, 1855	2a, 13, 15, 21, 23, 26, 29, 33, 35, 37, 38
Cerithium salebrosum Sowerby, 1855	2b, 6, 7, 13, 22, 23, 31, 33, 35, 41
Cerithium tenellum Sowerby, 1855	4, 8, 16
Cerithium tenuifilosum Sowerby, 1866	18
Cerithium trailli (Sowerby, 1855)	4, 9, 23, 27

SPECIES	SITE RECORDS
Cerithium zonatum (Wood, 1828)	38
Clypeomorus batillariaeformis Habe & Kosuge, 1966	4, 7, 8, 25
Clypeomorus moniliferum (Kiener, 1841)	4
Pseudovertagus aluco (Linnaeus, 1758)	15, 38
Rhinoclavis articulata (Adams & Reeve, 1850)	18, 25, 27, 28, 30, 37B, 38, 42, 44
Rhinoclavis aspera (Linnaeus, 1758)	2a, 3, 5, 7, 10-13, 15, 16, 18, 20-25, 27, 29, 30, 32, 33, 35, 38, 39, 41-43
Rhinoclavis fasciatus (Bruguière, 1792)	2b, 11, 18, 20, 22, 25, 37, 38, 43, 44
Rhinoclavis kochi (Philippi, 1848)	21
Rhinoclavis sinensis (Gmelin, 1791)	2b, 15, 18, 28, 31-33, 37, 39
Family Planaxidae	
Planaxis niger Quoy & Gaimard, 1834	27
Family Potamididae	
Cerithidea cingulata (Gmelin, 1791)	11
Terebralia sulcata (Born, 1778)	27
Telescopium telescopium (Linnaeus, 1758)	13
Family Modulidae	
Modulus tectum (Gmelin, 1791)	2b, 11, 27, 28, 30, 37, 44
Family Littorinidae	
Littoraria sp.	8, 9, 13, 19, 23, 44
Littorina undulata Gray, 1839	8, 15, 25, 28, 42, 44
Tectarius grandinatus (Gmelin, 1791)	13
Family Rissoinidae	
Rissoina reticulata (Sowerby, 1824)	11, 13
Family Rissoidae	
Zebina gigantea (Deshayes, 1850)	2b, 11
Family Strombidae	
Lambis chiragra (Linnaeus, 1758)	2a-3, 7, 31
Lambis lambis (Linnaeus, 1758)	12 ,13, 23, 33, 37, 40, 44
Lambis millepeda (Linnaeus, 1758)	2b, 3
Lambis scorpius (Linnaeus, 1758)	3, 7, 10, 13, 17, 26, 28, 36, 39, 44
Lambis truncatus (Humphrey, 1786)	2a, 19
Strombus bulla (Röding, 1798)	43
Strombus canarium Linnaeus, 1758	4
Strombus dentatus Linnaeus, 1758	2a, 2b, 22
Strombus gibberulus Linnaeus, 1758	2b, 3, 6, 11-13, 19, 20, 26, 29, 37B , 44
Strombus lentiginosus Linnaeus, 1758	3, 13, 26, 44
Strombus luhuanus Linnaeus, 1758	6, 7, 10, 11, 20, 24, 26, 30, 33, 35, 37, 38, 40, 43, 44
Strombus microurceus (Kira, 1959)	1-2b, 4, 6, 7, 10-13, 16, 21, 24-28, 30-34, 37-39
Strombus minimus Linnaeus, 1771	43
Strombus sinuatus Humphrey, 1786	2b
Strombus urseus Linnaeus, 1758	8, 12, 19, 27, 31, 44
Strombus variabilis Swainson, 1820	43
Terebellum terebellum (Linnaeus, 1758)	6-8, 10-12, 19, 22, 24, 27, 29, 31, 33, 35, 38

SPECIES	SITE RECORDS
Family Vanikoridae	
Vanikoro cancellata (Lamarck, 1822)	21, 22
Family Hipponicidae	
Hipponix conicus (Schumacher, 1817)	2b, 11, 15-17, 20, 27, 30, 32-34, 36, 43
Family Calyptraeidae	
Calyptraea extinctorium Lamarck, 1822	29
Family Capulidae	
Cheilea equestris (Linnaeus, 1758)	2b, 12, 13, 30, 34, 38, 39, 44
Family Vermetidae	
Serpulorbis colubrina (Röding, 1798)	2, 5, 6, 10, 12-14, 16-18, 20-24, 26, 30, 33, 34, 37-44
Vermetus cf. *tokyoensis* Pilsbry, 1895	8, 11, 13
Family Cypraeidae	
Cypraea annulus Linnaeus, 1758	17, 19, 26, 29, 44
Cypraea arabica Linnaeus, 1758	2a, 2b, 4, 5, 10, 14, 15, 17, 21, 22, 24, 28, 30, 32, 36, 37
Cypraea argus Linnaeus, 1758	2, 10, 36
Cypraea asellus Linnaeus, 1758	1, 2a, 6, 8, 11, 12, 14, 17, 21, 22, 24, 25, 31, 32, 34, 35, 42, 44
Cypraea becki Gaskoin, 1836	17, 42
Cypraea caputserpentis Linnaeus, 1758	1, 16B
Cypraea carneola Linnaeus, 1758	1, 2a, 3, 5, 7, 10-12, 15, 17, 18, 20, 21, 24, 28, 34, 36-39, 41
Cypraea chinensis Gmelin, 1791	1
Cypraea cicercula Linnaeus, 1758	7, 14, 42
Cypraea cribraria Linnaeus, 1758	1, 2a, 10, 17, 18, 28, 32, 33, 36, 39-41
Cypraea contaminata Sowerby, 1832	30
Cypraea cylindrica Born, 1778	1-2b, 5, 6, 8, 11, 13, 14, 17, 19, 20, 35, 38, 40, 42
Cypraea depressa Gray, 1824	34
Cypraea eglantina (Duclos, 1833)	39, 42
Cypraea erosa Linnaeus, 1758	2b ,8, 18, 21, 29, 31, 32, 35, 36-38, 40, 41, 44
Cypraea errones Linnaeus, 1758	4, 42
Cypraea felina Gmelin, 1791	1, 5
Cypraea fimbriata Gmelin, 1791	17, 18, 21, 22, 31, 34, 37, 39, 43
Cypraea gracilis Gaskoin, 1849	42
Cypraea globulus Linnaeus, 1758	38
Cypraea helvola Linnaeus, 1758	1, 7, 10, 16B, 21, 33
Cypraea isabella Linnaeus, 1758	1, 2a, 8, 9, 14-16, 21, 22, 25, 31, 32, 36, 39, 44
Cypraea kieneri Hidalgo, 1906	27
Cypraea labrolineata Gaskoin, 1848	7, 8, 13, 15, 18, 43
Cypraea limacina Lamarck, 1810	9, 12
Cypraea lutea Gmelin, 1791	30
Cypraea lynx Linnaeus, 1758	1, 2a, 3-5, 7, 8, 10-13, 15-17, 19, 21, 23, 25, 29, 30, 36, 38-40,43
Cypraea mappa Linnaeus, 1758	15
Cypraea mauritania Linnaeus, 1758	34
Cypraea minoridens Melvill, 1901	26

SPECIES	SITE RECORDS
Cypraea moneta Linnaeus, 1758	2b, 4, 6, 11-13, 15, 16, 20, 25, 26, 31, 34, 35, 38, 42
Cypraea nucleus Linnaeus, 1758	2, 3, 7, 11, 14, 16, 20, 22, 27, 44
Cypraea ovum Gmelin, 1791	4, 11
Cypraea pallidula Gaskoin, 1849	23
Cypraea punctata Linnaeus, 1758	27, 39
Cypraea quadrimaculata Gray, 1824	12, 13, 17, 38, 39, 40
Cypraea scurra Gmelin, 1791	32
Cypraea staphylaea Linnaeus, 1758	20, 28, 42
Cypraea talpa Linnaeus, 1758	7, 32
Cypraea teres Gmelin, 1791	2, 21, 26
Cypraea testudinaria Linnaeus, 1758	32
Cypraea tigris Linnaeus, 1758	4, 10, 13, 15, 18, 23, 41, 43, 44
Cypraea ursellus Gmelin, 1791	40
Cypraea vitellus Linnaeus, 1758	13, 15, 27
Family Ovulidae	
Calpurneus verrucosus (Linnaeus, 1758)	28, 32, 40
Diminovula punctata (Duclos, 1831)	20
Ovula ovum (Linnaeus, 1758)	3, 10, 43
Prionovolva brevis (Sowerby, 1828)	27, 30
Family Triviidae	
Trivia oryza (Lamarck, 1810)	1, 6, 11, 13, 17, 19, 28, 29, 33, 39, 41
Family Lamellariidae	
Coriocella nigra Blainville, 1824	27
Family Naticidae	
Natica gualteriana Récluz, 1844	33, 34, 37B
Natica onca (Röding, 1798)	39
Natica phytelephas Reeve, 1855	6
Natica violacea Sowerby, 1825	27, 30, 39, 42
Natica stellata Hedley, 1913	12
Natica vitellus (Linnaeus, 1758)	12
Polinices flemingianus (Récluz, 1844)	16B
Polinices aurantius (Röding, 1798)	
Polinices melanostomus (Gmelin, 1791)	6, 12, 13, 28, 39, 44
Polinices peselephanti (Link, 1807)	6, 7, 27, 42
Polinices tumidus (Swainson, 1840)	5, 7, 27, 32, 35, 37B, 38, 41
Family Bursidae	
Bursa cruentata (Sowerby, 1835)	31
Bursa granularis (Röding, 1798)	2 ,5, 13, 31, 32, 34, 37, 42
Bursa lamarckii (Deshayes, 1853)	2 ,13 ,34 ,41
Bursa leo Shikama, 1964	16B
Bursa rhodostoma Sowerby, 1840	7, 25, 27, 30, 34, 40
Bursa rosa Perry, 1811	28
Bursa tuberossissima (Reeve, 1844)	7, 14, 20
Tutufa rubeta (Linnaeus, 1758)	7, 11, 14, 16, 18, 20, 34, 39

SPECIES	SITE RECORDS
Family Cassidae	
Casmaria erinaceus (Linnaeus, 1758)	2b, 27, 34, 41-44
Cassis cornuta (Linnaeus, 1758)	5
Family Ranellidae	
Charonia tritonis (Linnaeus, 1758)	22, 26
Cymatium aquatile (Reeve, 1844)	16, 35
Cymatium flaveolum (Röding, 1798)	21
Cymatium hepaticum (Röding, 1798)	35
Cymatium lotorium (Linnaeus, 1758)	21, 33, 34
Cymatium mundum (Gould, 1849)	16B
Cymatium nicobaricum (Röding, 1798)	7, 11
Cymatium pileare (Linnaeus, 1758)	5, 35
Cymatium pyrum (Linnaeus, 1758)	3, 14, 20, 34
Cymatium rubeculum (Linnaeus, 1758)	1, 3, 13
Cymatium succinctum (Linnaeus, 1771)	2a
Distorsio anus (Linnaeus, 1758)	10, 28
Gyrineum cuspidatum (Reeve, 1844)	39
Gyrineum gyrinum (Linnaeus, 1758)	1, 6, 7, 10, 12, 13, 21, 35, 39, 41
Gyrineum pusillum (A. Adams, 1854)	8, 12, 39
Linatella succincta (Linnaeus, 1771)	41
Ranularia muricinum (Gmelin, 1791)	16B, 44
Septa gemmata (Reeve, 1844)	14 ,17
Septa vespacea (Lamarck, 1822)	27
Family Tonnidae	
Malea pomum (Linnaeus, 1758)	30, 41
Tonna allium (Dillwyn, 1817)	2b
Tonna cepa (Röding, 1798)	35
Tonna cumingii (Reeve, 1849)	19
Tonna perdix (Linnaeus, 1758)	2b, 11, 34
Family Cerithiopsidae	
Cerithiopsid sp. 1	39
Cerithiopsid sp. 2	39
Cerithiopsid sp. 3	41
Eulimidae	
Thyca crystallina (Gould, 1846)	7, 11, 13, 14, 21, 24, 35, 39
Family Muricidae	
Atilliosa nodulifera (Sowerby, 1841)	16, 31
Chicoreus brunneus (Link, 1810)	2b, 7, 15, 18, 20, 21, 30, 32, 37, 40
Chicoreus cumingii (A. Adams, 1853)	25
Chicoreus lacianatus (Sowerby, 1841)	27, 30
Chicoreus microphyllus (Lamarck, 1816)	7, 15, 21, 38
Chicoreus palmarosae (Lamarck, 1822)	39
Chicoreus strigatus (Reeve, 1849)	11
Coralliophila costularis (Lamarck, 1816)	2, 6, 11, 41
Coralliophila erosa (Röding, 1798)	18, 27

SPECIES	SITE RECORDS
Coralliophila neritoidea (Lamarck, 1816)	2a, 2b, 7-14, 16, 17, 20-22, 24-26, 29, 31-35, 37, 39-44
Cronia contracta (Reeve, 1846)	4
Cronia margariticola (Broderip, 1833)	8, 37B ,38
Drupa grossularia (Röding, 1798)	1-2b, 7, 13, 14, 16-18 ,22, 24, 32, 34, 37, 39, 41
Drupa morum (Röding, 1798)	1, 21, 26, 31
Drupa ricinus (Linnaeus, 1758)	1-2b, 10, 13, 16, 18, 24, 28, 30, 32, 33, 36, 37, 39, 41
Drupa rubusidaeus (Röding, 1798)	1-2b, 7, 15, 16, 21, 24, 26, 28, 30, 34, 37, 39
Drupella cariosa (Wood, 1828)	2a-4, 6, 12, 13, 16, 19, 23, 26, 27, 38, 40, 44
Drupella cornus (Röding, 1798)	1-2b, 11, 12, 15-17, 21, 22, 26, 28-30, 36, 37
Drupella ochrostoma (Blainville, 1832)	1-2b, 8, 11, 13, 15, 21, 22, 24, 26, 28, 31, 41, 43
Drupella rugosa (Born, 1778)	2, 11, 20
Favartia sp.	9, 34
Homalocantha scorpius (Linnaeus, 1758)	43
Maculotriton sculptile (Reeve, 1844)	33
Morula anaxeres (Kiener, 1835)	1-2b, 10, 14, 16, 17, 22, 25, 30, 32, 34, 39
Morula aurantiaca (Hombron & Jacquinot, 1853)	12, 19
Morula biconica (Blainville, 1832)	16
Morula dumosa (Conrad, 1837)	20, 38
Morula granulata (Duclos, 1832)	1, 4, 13, 16B, 18, 20, 23, 34, 39, 42, 44
Morula margariticola (Broderip, 1832)	5, 8, 23, 38
Morula musiva (Kiener, 1836)	4, 9, 40
Morula nodulifera (Menke, 1829)	33
Morula spinosa (H. & A. Adams, 1855)	1,2b, 11, 15-17, 27, 32, 35, 39, 42
Morula uva (Röding, 1798)	1, 2b, 3, 15, 16, 36, 41
Murex ramosus (Linnaeus, 1758)	5, 6
Naquetia triquetra (Born, 1778)	30-32
Nassa francolina (Bruguière, 1789)	43
Nassa serta (Bruguière, 1789)	33, 34
Pterynotus barclayanus (A. Adams, 1873)	16
Pterynotus tripterus (Born, 1778)	41
Quoyola madreporarum (Sowerby, 1832)	2a, 7, 10, 14, 17, 21, 22, 27, 32, 40, 43
Rapa rapa (Gmelin, 1791)	18, 21, 28, 39
Thais alouina Röding, 1798	4, 18, 36
Thais armigera (Link, 1807)	1, 20, 30, 32-34, 36
Thais echinata Blainville, 1832	15
Thais kieneri (Deshayes, 1844)	13, 18, 24, 39
Thais mancinella (Linnaeus, 1758)	1, 5, 7, 10, 14, 15, 21, 24, 27, 28, 30, 32
Thais savignyi (Deshayes, 1844)	13, 21
Thais tuberosa (Röding, 1798)	16, 20, 21, 44
Vitularia milaris (Gmelin, 1791)	5, 9, 12
Family Turbinellidae	
Vasum ceramicum (Linnaeus, 1758)	31
Vasum tubiferum (Anton, 1839)	32
Vasum turbinellus (Linnaeus, 1758)	2a, 3, 6, 13, 16, 19, 20, 24, 26, 30, 33, 34, 37, 39, 41

SPECIES	SITE RECORDS
Family Buccinidae	
Colubraria cf. *antiquata* (Hinds, 1844)	34
Colubraria castanea Kuroda & Habe, 1952	44
Colubraria nitidula (Sowerby, 1833)	2, 7, 14, 28, 34
Colubraria tortuosa (Reeve, 1844)	12
Cantharus iostomus (Gray in Griffith & Pidgeon, 1834)	18, 31, 38, 39
Cantharus pulcher (Reeve, 1846)	1, 3, 8, 10, 11, 15, 17, 25, 38, 43
Cantharus undosus (Linnaeus, 1758)	1-2b, 6, 7, 14, 15, 18, 20, 21, 29, 32, 34, 37, 38, 40
Cantharus wagneri (Anton, 1839)	3, 11, 13
Crassicantharus noumeensis (Crosse, 1870)	14
Engina alveolata (Kiener, 1836)	1, 38
Engina contracta (Reeve, 1846)	25, 38
Engina incarnata (Deshayes, 1834)	1, 11, 22, 30, 33, 39-42, 44
Engina lineata (Reeve, 1846)	1, 2b
Engina mendicaria (Linnaeus, 1758)	13, 37, 39, 44
Engina obliquicostata	27
Engina zonalis (Lamarck, 1822)	37B
Phos sculptilis Watson, 1886	28
Phos textum (Gmelin, 1791)	1-2b, 7, 10, 11, 14, 15, 19-22, 24-28, 30-33, 38, 39, 43, 44
Pisania gracilis (Reeve, 1846)	18, 31, 39
Caducifer decapitatus decapitatus (Reeve, 1844)	27
Family Columbellidae	
Mitrella albina (Kiener, 1841)	33, 38, 43
Mitrella ligula (Duclos, 1840)	2b, 5, 7, 27, 33, 34, 39, 42
Mitrella cf. *margarita* (Reeve, 1859)	10
Mitrella marquesa (Gaskoin, 1852)	33
Mitrella sp.	2b
Pyrene deshayesii (Crosse, 1959)	40
Pyrene flava (Bruguière, 1789)	7, 13, 21, 38
Pyrene livescens (Reeve, 1859)	27, 31, 34, 39
Pyrene ocellata (Link, 1807)	13, 37B, 39
Pyrene punctata (Bruguière, 1789)	1, 2b, 7, 10, 14, 16, 18, 21, 24, 25, 27, 30, 31, 33, 34, 36, 39, 41, 43, 44
Pyrene scripta (Lamarck, 1822)	1, 6, 34, 42
Pyrene testudinaria (Link, 1807)	2b, 21
Pyrene turturina (Lamarck, 1822)	7, 12, 13, 17, 19, 21, 22, 25, 27, 28, 31, 32, 35, 37, 39, 40, 43
Pyrene varians (Sowerby, 1832)	16, 33, 34
Family Nassariidae	
Hebra horrida (Dunker, 1847)	2b, 43
Nassarius acuticostus (Montrouzier, 1864)	7, 16B
Nassarius albescens (Dunker, 1846)	7, 33
Nassarius arcularius (Linnaeus, 1758)	7, 16B, 37B
Nassarius callospira (A. Adams, 1852)	7, 16B

SPECIES	SITE RECORDS
Nassarius comptus (A. Adams, 1852)	40
Nassarius concinnus (Powys, 1835)	13, 23, 29
Nassarius distortus (Adams, 1852)	7
Nassarius glans (Linnaeus, 1758)	33, 37
Nassarius granifer (Kiener, 1834)	27
Nassarius luridus (Gould, 1850)	16B
Nassarius margaratifer (Dunker, 1847)	19b
Nassarius pauperus (Gould, 1850)	7, 29, 42
Nassarius pullus (Linnaeus, 1758)	7, 27
Nassarius quadrasi (Hidalgo, 1904)	43
Nassarius reevianus (Dunker, 1847)	15,16B
Nassarius siquijorensis (A. Adams, 1852)	40
Nassarius venustus (Dunker, 1847)	16B
Family Fasciolariidae	
Dolicholatirus lancea (Gmelin, 1791)	1, 10, 20, 32, 38, 42
Latirolagena smaragdula (Linnaeus, 1758)	2b, 18, 24, 28, 30, 34
Latirus belcheri (Reeve, 1847)	3, 16
Latirus gibbulus (Gmelin, 1791)	22
Latirus lanceolatus (Reeve, 1847)	21, 22
Latirus nodatus (Gmelin, 1791)	1, 2, 5, 11, 14, 16-18, 20, 21, 28, 30, 34, 37
Latirus pictus (Reeve, 1847)	2b, 12, 22, 41-43
Latirus polygonus (Gmelin, 1791)	20
Latirus turritus (Gmelin, 1791)	1, 2b, 7, 14, 15, 17, 18, 28
Peristernia incarnata (Deshayes, 1830)	11
Peristernia nassatula (Lamarck, 1822)	16B, 22, 25, 30, 33, 36, 37
Peristernia ustulata (Reeve, 1847)	9, 15, 29, 40
Pleuroploca filamentosa (Röding, 1798)	1, 5, 18, 43
Pleuroploca trapezium (Linnaeus, 1758)	25
Family Volutidae	
Cymbiola aulica (Sowerby, 1825)	13, 19, 23
Cymbiola vespertilio (Linnaeus, 1758)	27 ,38
Melo sp	6
Family Olividae	
Oliva annulata (Gmelin, 1791)	2b, 6, 7, 11, 13-17, 20-22, 24, 25, 27, 28, 30-35, 37-39, 43
Oliva carneola (Gmelin, 1791)	2b, 7, 18, 20, 22, 29, 33, 35, 39, 42-44
Oliva lignaria Marrat, 1868	44
Oliva miniacea Röding, 1798	20
Oliva parkinsoni Prior, 1975	15, 18, 21, 26, 27, 30, 39, 41
Oliva tessellata Lamarck, 1811	2b, 28
Oliva tremulina Lamarck, 1810	20, 38, 44
Oliva tricolor Lamarck, 1811	44
Family Harpidae	
Harpa amouretta Röding, 1798	1, 10, 25

SPECIES	SITE RECORDS
Family Mitridae	
Cancilla fulgetrum (Reeve, 1844)	2
Cancilla gloriola Cernohorsky, 1970	21
Cancilla sp.	40
Imbricaria olivaeformis (Swainson, 1821)	2b, 10, 22, 23, 25-27, 33, 35, 37, 43
Imbricaria punctata (Swainson, 1821)	6, 16, 33
Imbricaria vanikorensis (Quoy & Gaimard, 1833)	6
Mitra acuminata Swainson, 1824	20, 33
Mitra aurantia (Gmelin, 1791)	10
Mitra coarctata Reeve, 1844	43
Mitra contracta Swainson, 1820	30, 32, 37-39
Mitra coronata Lamarck, 1811	41
Mitra cucumerina Lamarck, 1811	1, 13
Mitra decurtata Reeve, 1844	39
Mitra fraga (Quoy & Gaimard, 1833)	1, 6, 11, 22, 38, 42
Mitra imperialis Röding, 1798	5
Mitra litterata Lamarck, 1811	34
Mitra luctuosa A. Adams, 1853	15, 18, 28
Mitra lugubris Swainson, 1822	38
Mitra mitra (Linnaeus, 1758)	11, 16, 33
Mitra cf. *pelliserpentis* Reeve, 1844	20
Mitra punticulata Lamarck, 1811	5
Mitra retusa Lamarck, 1811	37, 39
Mitra rosacea Reeve, 1845	21, 38
Mitra scutulata (Gmelin, 1791)	18
Mitra stictica (Link, 1807)	14, 25
Mitra turgida Reeve, 1845	30
Mitra ustulata Reeve, 1844	30
Mitra vexillum Reeve, 1844	38
Neocancilla clathrus (Gmelin, 1791)	26
Neocancilla papilio (Link, 1807)	20, 30, 33, 41-43
Pterygia eximia (A. Adams, 1853)	33
Pterygia scabricula (Linnaeus, 1758)	25, 26
Scabricola desetangsii (Kiener, 1838)	33
Family Costellariidae	
Vexillum amanda (Reeve, 1845)	40
Vexillum cadaverosum (Reeve, 1844)	10
Vexillum citrinum (Gmelin, 1791)	9, 12
Vexillum consangiuneum (Reeve, 1844)	29
Vexillum costatum (Gmelin, 1791)	27
Vexillum crocatum (Lamarck, 1811)	2a, 21, 29
Vexillum curviliratum (Sowerby, 1874)	40
Vexillum dennisoni (Reeve, 1844)	9, 22
Vexillum deshayesii (Reeve, 1844)	27
Vexillum exasperatum (Gmelin, 1791)	2b, 12, 20, 35, 42, 44

SPECIES	SITE RECORDS
Vexillum granosum (Gmelin, 1791)	19, 29, 44
Vexillum microzonias (Lamarck, 1811)	36, 40
Vexillum lautum (Reeve, 1845)	16
Vexillum pacificum (Reeve, 1845)	7, 8, 10, 33, 37
Vexillum plicarium (Linnaeus, 1758)	7, 37B
Vexillum polygonum (Gmelin, 1791)	13, 26
Vexillum semicostatum (Anton, 1838)	5
Vexillum turrigerum (Reeve, 1845)	2b, 13, 20, 42
Vexillum vulpecula (Linnaeus, 1758)	23, 29
Family Turridae	
Clavus canalicularis (Röding, 1798)	12
Clavus exasperatus (Reeve, 1843)	13, 40
Clavus laetus (Hinds, 1843)	33
Clavus pica (Reeve, 1843)	3, 5
Clavus sp. 1	13
Clavus unizonalis (Lamarck, 1822)	7, 12, 33
Gemmula sp.	8
Inquisitor varicosa (Reeve, 1843)	43
Lienardia nigrocincta (Montrouzier, 1872)	33
Ptychobela flavidula (Lamarck, 1822)	37
Turridrupa astricta (Reeve, 1843)	26, 27
Turridrupa bijubata (Reeve, 1843)	21, 40
Turridrupa cerithina (Anton, 1839)	15, 16, 21, 25, 27, 28
Turridrupa sp.	39
Turris crispa (Lamarck, 1816)	40
Xenoturris cingulifera (Lamarck, 1822)	20, 27, 33, 35, 44
Family Terebridae	
Hastula lanceata (Linnaeus, 1767)	18, 32, 33, 38
Hastula penicillata (Hinds, 1844)	38
Hastula solida (Deshayes, 1857)	38
Hastula strigilata (Linnaeus, 1758)	25
Terebra affinis Gray, 1834	2b, 6, 10, 11, 16-18, 20-22, 27, 30, 33, 35, 37, 38, 43
Terebra areolata (Link, 1807)	2b, 12, 22, 42, 44
Terebra argus Hinds, 1844	20, 32, 38
Terebra babylonia Lamarck, 1822	2b, 6, 10, 11, 20, 22, 43, 44
Terebra caddeyi Bratcher & Cernohorsky, 1982	29, 43
Terebra cerithina Lamarck, 1822	15, 34
Terebra chlorata Lamarck, 1822	32, 38
Terebra cf. *circumcincta* Deshayes, 1857	38
Terebra cf. *columellaris* Hinds, 1844	26
Terebra crenulata (Linnaeus, 1758)	21
Terebra cumingi Deshayes, 1857	29
Terebra dimidiata (Linnaeus, 1758)	16B, 22
Terebra felina (Dillwyn, 1817)	7, 10, 14, 16B, 18, 30

SPECIES	SITE RECORDS
Terebra funiculata Hinds, 1844	2b, 10, 20, 38
Terebra guttata (Röding, 1798)	32
Terebra cf. *jenningsi* R.D. Burch, 1965	20
Terebra laevigata Gray, 1834	2b, 20, 22, 38, 39, 43, 44
Terebra maculata (Linnaeus, 1758)	2b, 13, 20, 22, 24
Terebra nebulosa Sowerby, 1825	7, 27, 30, 37
Terebra cf. *parkinsoni* Bratcher & Cernohorsky, 1976	11, 29
Terebra pertusa (Born, 1778)	25
Terebra solida Deshayes, 1857	32
Terebra undulata Gray, 1834	2b, 11, 22, 25, 43, 44
Terenolla pygmaea (Hinds, 1844)	30

Family Conidae

SPECIES	SITE RECORDS
Conus ammiralis Linnaeus, 1758	36
Conus arenatus Hwass in Bruguière, 1792	2b, 10-14, 16, 18, 20, 21, 24, 29-31, 33, 35, 38, 39, 41, 42,44
Conus auricomus Hwass in Bruguière, 1792	28
Conus balteatus Sowerby, 1833	30
Conus biliosus (Röding, 1798)	32
Conus canonicus Hwass in Bruguière, 1792	15
Conus capitaneus Linnaeus, 1758	2a, 6, 8, 9, 11, 12, 15, 26-30, 32-34, 37-39, 42
Conus catus Hwass in Bruguière, 1792	2b
Conus circumcisus Born, 1778	1
Conus consors Sowerby, 1833	43
Conus coronatus (Gmelin, 1791)	1, 2, 10, 14-16, 18, 28
Conus cylindraceus Broderip & Sowerby, 1833	41
Conus distans Hwass in Bruguière, 1792	2b, 31, 34-36
Conus ebraeus Linnaeus, 1758	2b, 15, 34
Conus flavidus Lamarck, 1810	1, 5, 7, 8, 30, 33, 34
Conus frigidus Reeve, 1843	26, 39
Conus geographus Linnaeus, 1758	18, 21, 28
Conus glans Hwass in Bruguière, 1792	17, 27
Conus imperialis Linnaeus, 1758	2a, 2b, 10, 14-16, 21, 26, 30-34, 43
Conus leopardus Röding, 1798	7, 10
Conus litoglyphus Röding, 1798	25, 26, 34, 38
Conus litteratus Linnaeus, 1758	2b, 7, 10, 13, 25, 26, 31, 32, 34, 37, 43, 44
Conus lividus Hwass in Bruguière, 1792	7, 15-17, 21, 30, 31, 33, 34, 36
Conus luteus Sowerby, 1833	15
Conus magus Linnaeus, 1758	3, 4, 5, 27, 37
Conus marmoreus Linnaeus, 1758	2b, 3, 13, 22, 40
Conus miles Linnaeus, 1758	1, 2a, 3, 6, 10, 11, 13-18, 20, 21, 24, 25, 28, 29, 33, 34, 37, 39, 43
Conus miliaris Hwass in Bruguière, 1792	2, 30, 33
Conus moreleti Crosse, 1858	1, 2a, 7, 12, 14, 18, 22, 28, 30, 31, 34, 36, 37
Conus mucronatus Reeve, 1843	16

SPECIES	SITE RECORDS
Conus muriculatus Sowerby, 1833	2a, 2b, 4, 5, 8, 15, 21, 26, 28, 32, 34
Conus musicus Hwass in Bruguière, 1792	1, 2a, 6, 7, 10, 11, 13-15, 17, 18, 20, 21, 24, 30, 32, 34, 38-40, 43
Conus mustelinus Hwass in Bruguière, 1792	12, 27, 30
Conus nussatella Linnaeus, 1758	38
Conus orbignyi Audouin, 1831	43
Conus parius Reeve, 1844	43
Conus planorbis Born, 1778	29, 40, 43
Conus praecellens A. Adams, 1854	43
Conus pulicarius Hwass in Bruguière, 1792	6, 7, 10, 16, 20, 22, 24, 29, 33, 39, 43
Conus rattus Hwass in Bruguière, 1792	1, 6, 13
Conus sanguinolentus Quoy & Gaimard, 1834	17, 18, 21, 22, 24, 28, 32
Conus spectrum Linnaeus, 1758	29, 37
Conus sponsalis Hwass in Bruguière, 1792	1, 2a, 21, 27, 28, 30, 33, 38, 40
Conus stercmuscarum Linnaeus, 1758	7, 16, 33, 34, 39
Conus striatellus Link, 1807	28, 30, 34
Conus striatus Linnaeus, 1758	2a, 21, 30, 43
Conus terebra Born, 1778	2, 14, 22, 30, 38, 39
Conus tessellatus Born, 1778	30, 43
Conus textile Linnaeus, 1758	2a, 2b, 7, 14, 15, 22, 25, 27-30, 32, 33, 43
Conus tulipa Linnaeus, 1758	44
Conus varius Linnaeus, 1758	12, 13, 15, 17, 18, 38
Conus vexillum Gmelin, 1791	2, 3, 10, 21, 22, 24, 31, 34, 38
Conus viola Cernohorsky, 1977	15, 32
Conus virgo Linnaeus, 1758	13, 44
Family Architectonicidae	
Heliacus dorsuosus (Hinds, 1844)	27
Philippia radiata (Röding, 1798)	43
Family Pyramidellidae	
Pyramidella sp.	2b
Family Aglajidae	
Chelidonura amoena Bergh, 1905	11, 12, 39
Chelidonure electra Rudman, 1970	35
Chelidonura inornata Baba, 1949	14
Philinopsis gardineri (Eliot, 1903)	3, 41
Family Haminoeidae	
Atys cylindricus (Helbling, 1779)	6, 12, 13, 42
Atys naucum (Linnaeus, 1758)	35, 40
Haminoea fusca (Pease, 1863)	40
Family Bullidae	
Bulla vernicosa Gould, 1859	16B
Family Plakobranchidae	
Plakobranchus ocellatus van Hasselt, 1824	3, 8, 11, 12, 16, 20, 21, 26, 27, 39, 40

SPECIES	SITE RECORDS
Family Elysiidae	
Elysia ornata (Swainson, 1840)	1 4, 27, 33
Elysia cf. *tomentosa* (Swainson, 1840)	1
Elysia sp.	34, 39
Thurdilla ratna Marcus, 1965	1, 37
Thurdilla sp.	34
Family Polyceridae	
Nembrotha kubaryana Bergh, 1877	12, 15, 28, 37, 43
Nembrotha aff. *lineolata* Bergh, 1905	11
Family Gymnodorididae	
Gymnodoris cf. *rubropapilosa* (Bergh, 1905)	8
Gymnodoris sp.	33
Family Dorididae	
Halgerda tessellata (Bergh, 1880)	3
Jorunna funebris (Kelaart, 1858)	5
Family Chromodorididae	
Chromodoris annae Bergh, 1877	33
Chromodoris coi Risbec, 1956	11, 15, 24, 32
Chromodoris collingwoodi Rudman, 1987	22
Chromodoris geometrica (Risbec, 1848)	5, 8
Chromodoris lochi Rudman, 1982	21
Glossodoris atromarginata (Cuvier, 1804)	34
Glossodoris cincta (Bergh, 1889)	20
Family Phyllidiidae	
Phyllidia coelestis Bergh, 1905	8, 15, 39
Phyllidia elegans Bergh, 1869	10, 13, 21, 39
Phyllidia ocellata (Cuvier, 1804)	8
Phyllidia pipeki Brunckhorst, 1993	8
Phyllidia varicosa Lamarck, 1801	3, 8, 18, 31, 39
Phyllidiella pustulosa (Cuvier, 1804)	1, 3, 5, 8, 12, 15, 16, 18, 21, 27, 28, 33, 36, 37, 44
Phyllidiopsis annae Brunckhorst, 1993	1
Phyllidiopsis sp.	1
Family Tethyidae	
Melibe fimbriata Alder & Hancock, 1864	35
Family Onchidiidae	
Onchidium sp.	37
Family Siphonariidae	
Siphonaria normalis (Gould, 1846)	34
Family Ellobiidae	
Cassidula sp.	39
Melampus fasciatus Deshayes, 1830	6, 32, 38
Pythia scabraeus (Linnaeus, 1758)	44
CLASS BIVALVIA	
Family Mytilidae	
Lithophaga sp.	1, 2b, 4-6, 8, 10, 12-18, 21-23, 25, 27, 29, 30, 32-39, 41-43

SPECIES	SITE RECORDS
Modiolus philippinarum Hanley, 1843	4, 10, 12, 19, 23, 27, 31, 35, 40, 43
Septifer bilocularis (Linnaeus, 1758)	1, 4, 8, 18, 19, 23, 27, 32, 34, 35, 38, 40
Septifer excisus (Wiegmann, 1837)	9, 12
Septifer sp.	15
Family Arcidae	
Anadara antiquata (Linnaeus, 1758)	2, 4, 9, 12, 13, 38, 44
Arca avellana (Lamarck, 1819)	2a-4, 7, 10-16, 18, 19, 21-24, 26, 27, 29, 30-35, 37-44
Arca navicularis Bruguière, 1798	1, 12, 35
Barbatia amygdalumtotsum (Röding, 1798)	1, 3, 4, 6, 8, 9, 11, 12, 15, 16, 19-21, 23-30, 38, 42, 43
Barbatia coma (Reeve, 1844)	5, 37B
Barbatia foliata Forskål, 1775	3-5, 18, 19, 21
Barbatia cf. *tenella* (Reeve, 1843)	5
Family Glycymerididae	
Glycymeris hoylei (Melvill & Standen, 1899)	9
Glycymeris reevei (Mayer, 1868)	2b, 15, 21, 43
Tucetona amboiensis (Gmelin, 1791)	2b, 4, 5, 9-11, 13, 15, 17, 22, 25, 26, 29, 32, 37, 44
Family Pteriidae	
Pinctada margaritifera (Linnaeus, 1758)	1, 2a, 5, 8, 9, 15-19, 21, 24, 30, 35-38, 44
Pinctada maxima (Jameson, 1901)	10, 14, 27, 35, 44
Pinctada nigra (Gould, 1850)	19
Pteria avicular (Holten, 1802)	5, 8, 21, 22, 33, 39, 42
Pteria pengiun (Röding, 1798)	5, 8, 12, 15, 16, 18, 21, 28, 39
Family Malleidae	
Malleus albus Lamarck, 1819	5
Malleus malleus (Linnaeus, 1758)	4, 5, 8, 29, 38
Vulsella vulsella (Linnaeus, 1758)	5, 19, 28, 29, 38
Family Isognomonidae	
Isognomon isognomum (Linnaeus, 1758)	2, 9, 15, 18, 23, 28, 29, 35, 38
Family Pinnidae	
Atrina vexillum (Born, 1778)	6, 9, 12, 27, 35, 40, 43
Pinna bicolor (Gmelin, 1791)	2b, 35, 38, 40
Streptopinna saccata (Linnaeus, 1758)	1, 11, 16, 18, 22, 31, 33, 39, 41, 43
Family Limidae	
Ctenoides ales (Finlay, 1927)	19
Lima cf. *basilanica* (A. Adams & Reeve, 1850)	40
Lima fragilis (Gmelin, 1791)	2, 3, 5, 27
Lima lima (Link, 1807)	3, 5-7, 9, 11, 12, 14-16, 18, 19, 21-24, 26-29, 31, 36-38, 40, 43
Lima orientalis (A. Adams & Reeve, 1850)	12-14, 18, 21, 23, 27, 29, 35, 38-41
Limatula japonica colmani Fleming, 1978	3
Family Ostreidae	
Alectryonella plicatula (Gmelin, 1791)	6, 11, 12, 17, 19, 20, 21, 23, 27, 29, 33, 40, 41, 43, 44
Dendostrea folium (Linnaeus, 1758)	6, 27, 41
Hyotissa hyotis (Linnaeus, 1758)	5, 6, 8, 10, 12, 15, 18, 27, 28, 35, 37-44

SPECIES	SITE RECORDS
Lopha cristagalli (Linnaeus, 1758)	5, 10, 14, 15, 41
Parahyotissa mumisma (Lamarck, 1819)	1, 2a, 3, 5-7, 10, 12, 14, 16-19, 21, 22, 24, 25, 32, 37, 40- 44
Saccostrea cf *cucullata* (Born, 1778)	1, 4-10, 18-21, 23, 24, 28, 30, 34, 36, 37, 40, 44
Saccostrea echinata (Quoy & Gaimard, 1835)	5, 21
Family Pectinidae	
Bratechlamys vexillum (Reeve, 1853)	11
Chlamys corsucans (Hinds, 1845)	1
Chlamys irregularis (Sowerby, 1842)	1
Decatopecten radula (Linnaeus, 1758)	1, 2b, 3-5, 7-12, 15, 16, 18-24, 25, 27-29, 31-35, 37, 40- 44
Exichlamys histronica (Gmelin, 1791)	8
Excellichlamys spectabilis (Reeve, 1853)	8
Gloripallium pallium (Linnaeus, 1758)	1, 2, 5, 7, 8, 10, 11, 13-16, 21, 22, 25, 27, 29-33, 35-44
Laevichlamys cf. *cuneata* (Reeve, 1853)	1, 10, 20, 22, 30, 39, 43
Laevichlamys deliciosa	16, 36, 38
Laevichlamys squamosa (Gmelin, 1791)	1, 5, 7-12, 20, 28, 35
Laevichlamys sp. 1	2b, 14
Mimachlamys punctata	1
Mimachlamys sp. 1	10
Mimachlamys sp. 2	2a, 28
Mirapecten moluccensis Dijkstra, 1988	11
Mirapecten rastellum (Lamarck, 1819)	39
Pedum spondyloidaeum (Gmelin, 1791)	1-5, 7, 8, 10-14, 16, 18-22, 24-26, 29, 31, 32, 35, 37, 39- 41, 44
Semipallium fulvicostatum (Adams & Reeve, 1850)	27, 29, 41, 42
Semipallium tigris (Lamarck, 1819)	7, 11, 12, 14, 42
Semipallium sp. 1	1, 12, 40
Semipallium sp. 2	1, 12
Family Spondylidae	
Spondylus barbatus Reeve, 1856	9
Spondylus butleri Reeve, 1856	5, 29
Spondylus candidus (Lamarck, 1819)	3, 7, 10, 11, 16, 21-23, 28-30, 33, 35, 40
Spondylus sanguineus Dunker, 1852	8, 14, 19, 23, 29
Spondylus sinensis Schreibers, 1793	1, 5, 7, 9-11, 13, 15-17, 22, 37-40, 44
Spondylus varians Sowerby, 1829	5, 6, 19
Family Chamidae	
Chama fibula Reeve, 1846	6, 11, 19
Chama lazarus Linnaeus, 1758	17, 29, 40
Chama limbula (Lamarck, 1819)	1-4, 6, 8, 12, 13, 15, 18, 19, 22, 26, 30, 31, 35, 37, 38, 40-44
Chama savigni Lamy, 1921	5, 38
Family Fimbriidae	
Codakia punctata (Linnaeus, 1758)	2b, 5, 6, 10, 20, 21, 23, 25, 27, 44
Codakia paytenorum (Iredale, 1937)	6
Codakia tigerina (Linnaeus, 1758)	12

SPECIES	SITE RECORDS
Fimbria fimbriata (Linnaeus, 1758)	12, 13, 22, 25, 29, 43, 44
Family Carditidae	
Beguina semiorbiculata (Linnaeus, 1758)	3, 7, 10, 12, 27, 35, 38, 40, 42, 44
Cardita variegata Bruguière, 1792	1, 3, 15, 28, 32, 36-39, 42, 44
Cardita sp.	12
Family Cardiidae	
Acrosterigma alternatum (Sowerby, 1841)	3-5, 8, 11, 12, 16, 19, 20, 22, 23, 27, 29-32, 35, 38-40, 42-44
Acrosterigma flava (Linnaeus, 1758)	6, 22, 40
Acrosterigma mindanaense (Sowerby, 1896)	1, 2a, 6, 8, 10, 12, 16, 19, 21, 23-25, 27-29, 33, 34, 36-43
Acrosterigma subrugosa (Sowerby, 1838)	9
Acrosterigma transcendens (Melvill & Standen, 1899)	13, 16, 27, 28, 37,38, 44
Acrosterigma unicolor (Sowerby, 1834)	5, 6, 21, 27, 39-41
Ctenocardia fornicata (Sowerby, 1840)	40
Fragum unedo (Linnaeus, 1758)	3, 13, 27, 29, 37B, 38, 42, 44
Fulvia australe (Sowerby, 1834)	4, 11, 21, 26, 40, 43
Laevicardium attenuatum (Sowerby, 1840)	5, 7, 11, 13, 18, 27-29, 33, 35, 39, 41-44
Laevicardium biradiatum (Bruguière, 1789)	12, 25, 28, 29
Laevicardium multipunctatum (Sowerby, 1833)	7, 23
Lyrocardium lyratum (Sowerby, 1840)	8, 9
Trachycardium enode (Sowerby, 1841)	2a, 5, 7, 8, 11, 14, 27, 38
Family Tridacnidae	
Hippopus hippopus (Linnaeus, 1758)	2b, 3, 11, 15, 16, 23, 26, 35, 42, 44
Hippopus porcellanus Rosewater, 1982	25, 29, 31
Tridacnea crocea Lamarck, 1819	1-3, 6, 7, 10-13, 16, 19-23, 25-27, 30, 31, 34, 35, 37-44
Tridacna derasa (Röding, 1798)	11, 12, 17, 19, 26, 27, 38
Tridacna gigas (Linnaeus, 1758)	6, 12, 22, 31
Tridacna maxima (Röding, 1798)	15, 16, 25, 26, 31
Tridacna squamosa Lamarck, 1819	2b, 3, 5-,7 ,9-11, 13, 16-20, 22, 24-27, 29-35, 37-44
Family Solenidae	
Solen sp.	9
Family Tellinidae	
Tellina bougei Sowerby, 1909	44
Tellina exculta Gould, 1850	10
Tellina gargadia Linnaeus, 1758	44
Tellina linguafelis Linnaeus, 1758	2b, 4, 16, 19, 27
Tellina palatum (Iredale, 1929)	12
Tellina rastellum Hanley, 1844	27
Tellina rostrata Linnaeus, 1758	42
Tellina scobinata Linnaeus, 1758	2, 39, 44
Tellina staurella Lamarck, 1818	23, 27
Tellina cf *tenuilamellata* Smith, 1885	3
Tellina umbonella Lamarck, 1818	27

SPECIES	SITE RECORDS
Family Mactridae	
Lutraria australis Reeve, 1854	5, 28, 32
Family Semelidae	
Semele casta A. Adams, 1853	12, 29
Semele duplicata (Sowerby, 1833)	27, 37, 40
Semele lamellosa (Sowerby, 1830)	41, 44
Family Psammobiidae	
Gari amethystus (Wood, 1815)	1, 3, 5, 8-10, 12, 14-16, 18, 26, 27, 29, 33, 37, 39, 43
Gari maculosa (Lamarck, 1818)	5, 40
Gari occidens (Gmelin, 1791)	3, 5, 11
Gari oriens (Deshayes, 1855)	21, 35
Gari pulcherrimus (Deshayes, 1855)	8, 18, 27, 28
Gari squamosa (Lamarck, 1818)	27
Family Trapeziidae	
Trapezium bicarinatum (Schumacher, 1817)	1, 38-40
Trapezium obesa (Reeve, 1843)	9, 19, 23
Family Veneridae	
Antigona chemnitzii (Hanley, 1844)	1, 5, 9, 15, 22, 40-42
Antigona aff. *persimilis* (Iredale, 1930)	10, 11, 15, 20, 21, 23, 30
Antigona purpurea (Linnaeus, 1771)	3, 5, 12, 14-16, 18, 21, 25, 28, 44
Antigona restriculata (Sowerby, 1853)	2a-12, 14, 15, 18, 20-22, 24, 25, 27, 29-32, 35-37, 39, 40, 42, 43
Antigona reticulata (Linnaeus, 1758)	2b, 11, 21, 30, 32, 34
Callista impar (Lamarck, 1818)	19
Circe scripta (Linnaeus, 1758)	1
Dosinia aff. *histrio* (Gmelin, 1791)	2b, 27
Dosinia iwakawai Oyama & Habe, 1 970	10, 11, 28, 29, 34, 35, 37, 40
Dosinia sp.	3, 5-7, 11, 15, 18, 21, 22, 27, 43
Globivenus capricorneus (Hedley, 1908)	27
Globivenus toreuma (Gould, 1850)	1, 2a, 3, 5, 7, 10, 13, 14, 16, 17, 20-22, 24, 26-29, 31-34, 36-39, 41-44
Katelysia ferruginea (Reeve, 1864)	19
Katelysia hiantina (Lamarck, 1818)	40
Lioconcha castrensis (Linnaeus, 1758)	2b, 5, 10-12, 18, 20, 22, 27, 35, 38, 42, 44
Lioconcha fastigiata (Sowerby, 1851)	23, 42
Lioconcha ornata (Dillwyn, 1817)	1, 3-5, 7-9, 11-13, 15-17, 19, 20, 27, 29-31, 37-40, 43, 44
Meretrix meretrix (Linnaeus, 1758)	13
Pitar spoori Lamprell & Whitehead, 1990	19, 27
Pitar sp.	19, 38
Samarangia quadrangularis (Adams & Reeve, 1850)	42
Sunetta sp.	39
Tapes literatus (Linnaeus, 1758)	3-5, 12, 15-17, 29, 38, 40-43
Tapes platyptycha Pilsbry, 1901	12, 15, 16, 21, 29, 40
Tapes sericus Matsukuma,1986	39
Tapes sulcarius Lamarck, 1818	38

SPECIES	SITE RECORDS
Tawera lagopus (Lamarck, 1818)	11, 37-39
Timoclea costillifera (Adams & Reeve, 1850)	11, 29, 32, 39
Family Corbulidae	
Corbula cf *tahitensis* Lamarck, 1818	5, 6, 14, 18, 21, 22, 28, 30-32, 35, 43
Family Pholadidae	
Martesia multistriata (Sowerby, 1849)	15
Family Clavagellidae	
Brechites sp.	3
Family Corbiculidae	
Glauconome rugosa Reeve, 1844	5, 12, 16
CLASS CEPHALOPODA	
Family Spirulidae	
Spirula spirula (Linnaeus, 1758)	On beaches
Family Sepiidae	
Sepia latimanus Quoy & Gaimard, 1832	13
Sepia papuaensis Hoyle, 1885	13-16
Family Loliginidae	
Loliginid sp.	13
Family Octopodidae	
Octopus sp.	2a
CLASS SCAPHOPODA	
Family Dentaliidae	
Dentalium aprinum Linnaeus, 1766	43
Dentalium elephantinum Linnaeus, 1758	12, 16, 29, 31, 40, 42, 43

Appendix 4

List of the reef fishes of the Raja Ampat Islands

G.R. Allen

This list includes all species of shallow (to 50 m depth) coral reef fishes known from the Raja Ampat Islands at 1 May 2001.

The list is based on the following sources:
1) Results of the 2001 CI Marine RAP; 2) 26 hours of scuba-diving observations by G. Allen in 1998 and 1999.

The numbers under the site records column and remarks in the abundance column pertain to the 2000 survey.

The phylogenetic sequence of the families appearing in this list follows Eschmeyer (Catalog of Fishes, California Academy of Sciences, 1998) with slight modification (eg. placement of Cirrhitidae). Genera and species are arranged alphabetically within each family. An asterisk (*) appearing after the author name(s) indicates that the species was previously recorded from the Raja Ampat Islands in published literature.

Terms relating to relative abundance are as follows: *Abundant* - Common at most sites in a variety of habitats with up to several hundred individuals being routinely observed on each dive. *Common* - seen at the majority of sites in numbers that are relatively high in relation to other members of a particular family, especially if a large family is involved. *Moderately common* - not necessarily seen on most dives, but may be relatively common when the correct habitat conditions are encountered. *Occasional* - infrequently sighted and usually in small numbers, but may be relatively common in a very limited habitat. *Rare* - less than 10, often only one or two individuals seen on all dives

SPECIES	SITE RECORDS	ABUNDANCE	DEPTH (m)
ORECTOLOBIDAE			
Eucrossorhinus dasypogon (Bleeker, 1867)*	14, 15, 16, 33	Rare, only 4 seen. Photographed. Waigeo is type locality.	4-12 m
HEMISCYLLIIDAE			
Hemiscyllium freycineti (Quoy & Gaimard, 1824)*	13	Rarely seen, but nocturnal. Waigeo is type locality. Photographed	1-15
GINGLYMOSTOMATIDAE			
Nebrius ferrugineus (Lesson, 1830)*	14	Rare, a single individual recorded. Waigeo is type locality.	1-70
CARCHARHINIDAE			
Carcharhinus amblyrhynchos (Bleeker, 1856)	1998-99		0-100
C. melanopterus (Quoy and Gaimard, 1824)*	1, 20, 31	Rare, less than five individuals observed. Waigeo is type locality.	0-10
Triaenodon obesus (Rüppell, 1835)*	2a	Rare, only one seen.	2-100
DASYATIDIDAE			
Dasyatis kuhlii (Müller and Henle, 1841)	8, 9, 12, 14-16, 18, 24, 29, 30	Occasionally seen in sandy areas.	2-50
Taeniura lymma (Forsskål, 1775)*	2b, 6, 7, 15, 16, 26, 39, 44	Occasionally seen in sandy areas.	2-30
MYLIOBATIDAE			
Aetobatus narinari (Euphrasen, 1790)	1998-99		0-25
MOBULIDAE			
Manta birostris (Walbaum, 1792)	28	Only one seen at sites, but about 30 observed near Weh Island.	0-100
Mobula tarapacana (Philippi, 1892)	1998-99		0-30
MORINGUIDAE			
Moringua ferruginea (Bliss, 1883)*	1998-99		1-10
M. microchir Bleeker, 1853	2b	Two specimens collected with rotenon.	3-20
MURAENIDAE			
Echidna nebulosa (Thünberg, 1789)*	1998-99		1-10
Gymnothorax enigmaticus McCosker and Randall, 1982	1998-99		5-40
G. fimbriatus (Bennett, 1831)	1998-99		0-30
G. flavimarginatus (Rüppell, 1828)	1998-99		1-150
G. fuscomaculatus (Schultz, 1953)	6, 41, 42		3-25
G. javanicus (Bleeker, 1865)	14, 36, 43	Rare, only 3 seen.	0.5-50

SPECIES	SITE RECORDS	ABUNDANCE	DEPTH (m)
G. melatremus Schultz, 1953	20	A single specimen collected with rotenone.	5-30
G. pictus (Ahl, 1789)*	31	One seen on shallow reef flat.	1-40
G. polyuranodon (Bleeker, 1853)	1998-99		0-3
G. zonipectus Seale, 1906	20, 41, 42	Three specimens collected with rotenone.	8-45
Rhinomuraena quaesita Garman, 1888	1998-99		1-50
Uropterygius micropterus (Bleeker, 1852)*	1998-99		1-20
OPHICHTHIDAE			
Leiuranus semicinctus (Lay and Bennett, 1839)*	1998-99		0-20
Myrichthys colubrinus (Boddaert,1781)	13	Only one seen, but cryptic and nocturnal.	0-8
M. maculosus (Cuvier, 1817)	2b	Several specimens collected with rotenone.	5-25
CONGRIDAE			
Gorgasia maculata Klausewitz & Eibesfeldt, 1959	10, 43	Large colony containing hundreds of individuals at site 43.	20-50
Heteroconger haasi (Klausewitz and Eibl-Eibesfeldt, 1959)	25	One colony observed.	3-45
CLUPEIDAE			
Herklotsichthys quadrimaculatus (Rüppell, 1837)*	3	One school seen at site 3.	0-3
Spratelloides delicatulus (Bennett, 1832)	6, 11-13, 27, 33, 41, 42, 44	Occasional, hundreds seen schooling near surface at several sites.	0-1
S. lewisi Wongratana, 1983	38	Several shoals seen at site 38. Photographed.	0-10
PLOTOSIDAE			
Plotosus lineatus (Thünberg, 1787)*	25, 26, 29, 30	Occasional.	1-20
SYNODONTIDAE			
Saurida gracilis (Quoy & Gaimard, 1824)	3, 38	Occasional on sand bottoms.	1-30
S. nebulosa Valenciennes, 1849	1998-99		1-30
Synodus dermatogenys Fowler, 1912	1-3, 7, 10, 12-15, 18, 21, 25, 26, 29-31, 33, 34, 38, 42	Moderately common, solitary individuals usually seen resting on dead coral or rubble. Photographed.	1-25
S. jaculum Russell and Cressy, 1979	7, 10, 12, 18, 25	Occasional on rubble bottoms. Photographed.	10-50

SPECIES	SITE RECORDS	ABUNDANCE	DEPTH (m)
S. rubromarmoratus Russell and Cressy, 1979	10, 12, 15	Rare, on sand or rubble bottoms.	5-30
S. variegatus (Lacepède, 1803)*	17, 18, 42	Occasional, solitary individuals or pairs usually seen resting on live coral. Photoraphed.	5-50
CARAPIDAE			
Onuxodon margaritiferae (Rendahl, 1921)	1998-99		2-30
BYTHITIDAE			
Brosmophyciops pautzkei Schultz, 1960	41	One specimen collected with rotenone.	5-55
Ogilbia sp. 1	41	Collected with rotenone.	
Ogilbia sp. 2	1998-99		0-5
BATRACHOIDIDAE			
Batrachomoeus trispinosus (Günther, 1861)	1998-99		
Halophryne diemensis (La Sueur, 1834)*	1998-99		
ANTENNARIIDAE			
Antennarius coccineus (Lesson, 1829)	1998-99		1-40
A. dorhensis Bleeker, 1859	1998-99		1-15
Histiophryne cryptacanthus (Weber, 1913)	2b	One specimen collected and photographed.	2-30
Histrio histrio (Linnaeus, 1758)*	33	Common in floating Sargassum.	0-1
GOBIESOCIDAE			
Diademichthys lineatus (Sauvage, 1883)	2a, 3, 5, 7, 12, 16, 25, 27-30, 37, 43, 44	Occasional among sea urchins or branching coral.	3-20
Discotrema crinophila Briggs, 1976	1998-99		5-30
ATHERINIDAE			
Atherinomorus lacunosus (Forster, 1801)	4, 8, 9, 12, 27, 41	Occasional large shoals seen.	0-2
Hypoatherina temminckii (Bleeker, 1853)	13	Locally abundant at one site. Collected.	0-2
BELONIDAE			
Platybelone platyura (Bennett, 1832)	25, 26	Rare.	0-2

SPECIES	SITE RECORDS	ABUNDANCE	DEPTH (m)
Tylosurus crocodilus (Peron & Lesueur, 1821)	1, 6, 11, 26, 42	Occasional, on surfaces at several sites.	0-4
HEMIRAMPHIDAE			
Hemirhamphus far (Forsskål, 1775)	13	Rare, only one seen.	0-2
Hyporhamphus. dussumieri (Valenciennes, 1846)	1998-99		0-2
Zenarchopterus buffonis (Valenciennes, 1847)	1998-99		0-2
Z. dunckeri Mohr, 1926	4, 8, 23, 27, 31	Common near mangroves.	0-2
HOLOCENTRIDAE			
Myripristis adusta Bleeker, 1853	2a, 7, 10, 14, 17, 21, 23, 28, 36, 39	Occasional, sheltering in caves and under ledges. Common at site 39.	3-30
M. berndti Jordan and Evermann, 1902	18, 21, 22, 25, 30, 36	Occasional, sheltering in caves and under ledges. Common at site 25.	8-55
M. botche Cuvier, 1829	2a, 14	Rare, only 5 fish seen. Photographed.	12-240
M. hexagona (Lacepède, 1802)	1, 2b, 4, 5, 8, 9, 11, 12, 14-16, 19, 26, 29, 31, 38, 40-42	Common, usually in coastal areas affected by silt.	10-40
M. kuntee Valenciennes, 1831	1, 2a, 7, 10, 13, 14, 16, 17, 21, 22, 25, 26, 28, 30-34, 36, 42	Moderately common, sheltering in caves and under ledges, but frequently exposes itself for brief periods.	5-30
M. murdjan (Forsskål, 1775)	3, 21, 25, 32, 33, 42	Occasional, sheltering in caves and under ledges.	3-40
M. pralinia Cuvier, 1829	25, 32	Only a few seen, but nocturnal.	3-40
M. violacea Bleeker, 1851	2a, 2b, 3, 6, 7, 10, 11, 13-19, 21-23, 25, 26, 30-37, 39, 40, 43, 44	Common, most abundant squirrelfish seen in Raja Ampats.	3-30
M. vittata Valenciennes, 1831	36	Rare. Seen at only one site.	12-80
Neoniphon opercularis (Valenciennes, 1831)	30, 32	Rare, only 2 seen.	3-20
N. sammara (Forsskål, 1775)*	2a, 2b, 3, 6, 11, 13, 25, 30, 37, 43, 44	Occasional, usually among branches of staghorn *Acropora* coral.	2-50
Sargocentron caudimaculatum (Rüppell, 1835)*	1, 2a, 7, 10, 14, 17, 18, 21, 22, 25, 26, 28, 30-34, 36, 37, 39, 41	Moderately common.	6-45
S. cornutum (Bleeker, 1853)	7	Rare, only 3 seen.	6-50

SPECIES	SITE RECORDS	ABUNDANCE	DEPTH (m)
S. diadema (Lacepède, 1802)	2b, 7, 13, 22, 31		2-30
S. melanospilos (Bleeker, 1858)	18, 33	Rare, only a few seen at two sites.	10-25
S. rubrum (Forsskål, 1775)*	4, 5, 7-9, 15, 16, 35	Occasional.	5-25
S. spiniferum (Forsskål, 1775)	2a, 3, 5, 9-12, 14, 17, 23, 26, 29, 33, 36, 37, 41	Moderately common, in caves and under ledges.	5-122
S. tiere (Cuvier, 1829)	31	Rare, but nocturnal.	10-180
S. violaceum (Bleeker, 1853)*	13, 32, 35, 42	Occasional.	3-30
PEGASIDAE			
Eurypegasus draconis (Linnaeus, 1766)	1998-99		2-20
AULOSTOMIDAE			
Aulostomus chinensis (Linnaeus, 1766)*	2a, 2b, 15, 16, 18, 26. 33, 37, 44	Occasional.	2-122
FISTULARIIDAE			
Fistularia commersoni Rüppell, 1835*	17, 31, 40, 43	Occasional.	2-128
CENTRISCIDAE			
Aeoliscus strigatus (Günther, 1860)*	1998-99		1-30
Centriscus scutatus (Linnaeus, 1758)*	34	Rare, only one aggregation seen.	1-30
SYNGNATHIDAE			
Choeroichthys brachysoma Bleeker, 1855	32	Collected with rotenone.	1-25
Corythoichthys flavofasciatus (Rüppell, 1838)	21	Rare, only one seen.	1-25
C. haematopterus (Bleeker, 1851)*	1998-99		1-20
C. intestinalis (Ramsay, 1881)	29, 44	Only seen at two sites and in low numbers.	1-25
C. ocellatus Herald, 1953	38	Rare. Photographed.	1-15
C. schultzi Herald, 1953	38	Rare.	1-30
C. sp. 1	28, 41	Rare.	1-25
C. sp. 2	28	Rare.	1-20
Doryrhamphus dactyliophorus (Bleeker, 1853)	12, 23	Only two seen, but a secretive cave and ledge dweller.	1-56

SPECIES	SITE RECORDS	ABUNDANCE	DEPTH (m)
D. janssi (Herald & Randall, 1972)	1998-99		5-35
Halicampus dunckeri (Kaup, 1856)	29, 42	Collected with rotenone.	2-14
Halicampus mataafae (Jordan & Seale, 1906)	1998-99		3-25
Hippocampus bargibanti Whitley, 1970	1998-99		10-40
H. kuda Bleeker, 1852	1998-99		0-12
Phoxocampus belcheri (Kaup, 1856)	1998-99		1-10
P. tetrophthalmus (Bleeker, 1858)	1998-99		1-10
Siokunichthys nigrolineatus Dawson, 1983	1998-99		10-20
Syngnathoides biaculeatus (Bloch, 1785)*	1998-99		0-10
SCORPAENIDAE			
Dendrochirus zebra (Cuvier, 1829)*	1998-99		1-20
Pterois antennata (Bloch, 1787)*	15, 17, 20	Rare, but mainly nocturnal. Photographed.	1-50
P. volitans (Linnaeus, 1758)	30, 31, 33	Rare.	2-50
Scorpaenodes guamensis (Quoy and Gaimard, 1824)*	1998-99		0-10
S. hirsutus (Smith, 1957)	32	Two collected with rotenone.	5-40
S. parvipinnis (Garrett, 1863)	32	One collected with rotenone.	2-50
Scorpaenopsis macrochir Ogilby, 1910	1998-99		1-10
S. oxycephala (Bleeker, 1849)	12	Rare, only one seen.	1-40
Sebastapistes cyanostigma (Bleeker, 1856)	20	Probably not uncommon, but only one seen among coral branches.	2-15
S. strongia (Cuvier, 1829)	1998-99		1-15
Taenianotus triacanthus Lacepède, 1802	1998-99		5-130
TETRAROGIDAE			
Ablabys macracanthus (Bleeker, 1852)	Photographed by R.Steene 2001		1-15
SYNANCEIIDAE			
Inimicus didactylus (Pallas, 1769)*	13, 41	Rare, only 2 seen.	5-40
Synanceja horrida (Linnaeus, 1766)	1998-99		0-10

SPECIES	SITE RECORDS	ABUNDANCE	DEPTH (m)
Synanceia verrucosa (Bloch & Schneider, 1801)*	1998-99		0-20
CARACANTHIDAE			
Caracanthus maculatus (Gray, 1831)	7		3-15
DACTYLOPTERIDAE			
Dactyloptena orientalis (Cuvier, 1829)*	1998-99		1-45
PLATYCEPHALIDAE			
Cociella punctata (Cuvier, 1829)	1998-99		1-20
Cymbacephalus beauforti Knapp, 1973	12	Only one seen, but difficult to detect.	2-12
Platycephalus sp.	1998-99		1-15
Thysanophrys chiltoni Schultz, 1966	28	One collected with rotenone.	1-80
CENTROPOMIDAE			
Psammoperca waigiensis (Cuvier, 1828)*	6	Rare, only one seen. Waigeo is type locality.	1-20
SERRANIDAE			
Aethaloperca rogaa (Forsskål, 1775)	2, 3, 6, 14, 17, 26, 30, 32, 37, 39	Occasional.	1-55
Anyperodon leucogrammicus (Valenciennes, 1828)	10, 15, 16, 33, 35	Occasional	5-50
Cephalopholis argus Bloch and Schneider, 1801	1, 2a, 7, 24, 26, 30, 31, 37, 44	Occasional.	1-40
C. boenack (Bloch, 1790)*	1, 4, 5, 8, 12, 16, 19, 23, 26, 27, 29, 35, 37, 38, 40-42, 44	Moderately common.	1-20
C. cyanostigma (Kuhl and Van Hasselt, 1828)	2b, 4-6, 8-13, 15-18, 20-26, 28, 29, 31, 33, 35, 37, 38, 41, 43, 44	Moderately common on sheltered reefs.	2-35
C. leopardus (Lacepède, 1802)	1-3, 7, 10, 13, 21, 22, 25, 26, 31, 32, 36, 44	Occasional.	3-25
C. microprion (Bleeker, 1852)	2b, 6, 9, 11, 16, 23, 29, 35, 40, 41, 43, 44	Occasional on relatively silty reefs.	2-20

SPECIES	SITE RECORDS	ABUNDANCE	DEPTH (m)
C. miniata (Forsskål, 1775)*	1-3, 7, 10, 14, 15, 17, 18, 20-22, 24, 25, 28, 30, 31, 33, 37, 39, 44	Moderately common, usually in areas of clear water.	3-150
C. sexmaculata Rüppell, 1828	10, 15, 20, 21, 36	Occasional, on ceilings of caves on steep drop-offs.	6-140
C. sonnerati (Valenciennes, 1828)	6, 42, 43	Rare, only 3 seen.	10-100
C. spiloparaea (Valenciennes, 1828)	13, 18, 22, 31, 33, 36	Occasional in deep water (below 20 m) of outer slopes.	16-108
C. urodeta (Schneider, 1801)*	1-3, 7, 10, 13-15, 18, 20, 24-26, 28, 31-34, 36, 37, 39, 42-44	Moderately common in variety of habitats.	1-36
Cromileptes altivelis (Valenciennes, 1828)	1, 4, 11	Rare, only 3 seen.	2-40
Diploprion bifasciatum Cuvier, 1828	3, 7, 11, 13, 15, 16, 18, 21-26, 28, 29, 31, 33, 34, 35, 39	Occasional, sheltered inshore areas.	2-25
Epinephelus areolatus (Forsskål, 1775)	1998-99		6-200
E. bilobatus Randall & Allen, 1987	1998-99		2-30
E. caeruleopunctatus (Bloch, 1790)*	39	Rare, only one adult seen.	5-25
E. coioides (Hamilton, 1822)	1998-99		2-100
E. corallicola (Kuhl and Van Hasselt, 1828)	14	Rare, only one seen.	3-15
E. fasciatus (Forsskål, 1775)	1, 4-6, 10, 15, 17, 18, 22, 24, 26, 28, 30, 32, 34, 36, 39, 44	Moderately common.	4-160
E. fuscoguttatus (Forsskål. 1775)*	2a, 14, 39	Rare, only 3 seen.	3-60
E. lanceolatus (Bloch, 1790)	14	Rare, only one seen.	3-10
E. macrospilos (Bleeker)	24	Rare, only one seen.	5-25
E. maculatus (Bloch, 1790)	19	Rare, only one seen.	10-80
E. merra Bloch, 1793	13, 16, 31, 38, 40, 44	Occasional, but common at site 38.	1-15
E. ongus (Bloch, 1790)	29, 32, 36, 40	Rare, only 4 seen.	5-25
E. polyphekadion (Bleeker, 1849)	39	Rare, only one seen.	2-45
E. spilotoceps Schultz, 1953	30	Rare, only one seen.	1-15

SPECIES	SITE RECORDS	ABUNDANCE	DEPTH (m)
E. tukula Morgans, 1959	10	Rare, only one seen.	1-15
Gracila albimarginata (Fowler and Bean, 1930)	1, 10, 32	Rare, only 3 seen.	6-120
Grammistes sexlineatus (Thünberg, 1792)*	1998-99		0.5-30
Grammistops ocellatus Schultz, 1953	41	Two specimens collected with rotenone.	5-30
Liopropoma susumi (Jordan & Seale, 1906)	32	One specimen collected with rotenone.	2-34
Luzonichthys waitei (Fowler, 1931)	1, 28, 43	Three large aggregations seen.	10-55
Plectropomus areolatus (Rüppell, 1830)	19, 21, 26	Rare, only three seen.	2-30
P. laevis (Lacepède, 1802)	1998-99		4-90
P. leopardus (Lacepède, 1802)	1, 6, 23, 27	Rare, about 8 seen.	3-100
P. maculatus (Bloch, 1790)	4, 9, 11, 15, 17, 26, 40	Occasional in silty areas.	2-30
P. oligocanthus (Bleeker, 1854)	1, 16, 17, 26, 32, 35, 37, 39, 42-44	Occasional.	4-40
Pogonoperca punctata (Valenciennes, 1830)	1998-99		5-40
Pseudanthias dispar (Herre, 1955)	2a, 7, 10, 17, 25, 26, 28, 39	Moderately common and locally abundant, but seen at few sites.	4-40
P. fasciatus (Kamohara, 1954)	21, 25	Rare, about 6 seen.	20-150
P. huchtii (Bleeker, 1857)	1-3, 6, 7, 9-18, 20-26, 28, 30-34, 36, 37, 39, 41-44	Abundant, one of most common reef fishes at Raja Ampats. Especially common at site 18.	4-20
P. hypselosoma Bleeker, 1878*	25, 29, 44	Rare, only a few seen at three sites.	10-40
P. luzonensis (Katayama and Masuda, 1983)	33	Several seen below 30 m depth.	12-60
P. pleurotaenia (Bleeker, 1857)	1, 2a, 16, 25, 31, 36	Occasional.	15-180
P. randalli (Lubbock & Allen, 1978)	20	Rare, 2 collected with rotenone.	15-70
P. squamipinnis (Peters, 1855)	1, 2a, 7, 14, 15, 17, 18, 21, 22, 25, 26, 28, 30, 32, 33, 36, 39, 43	Common, but usually less abundant than the similar P. huchtii.	4-20
P. tuka (Herre and Montalban, 1927)	2a, 7, 11, 13, 18, 20-22, 25, 26, 32, 43, 44	Occasional	8-25
Pseudogramma polyacanthum (Bleeker, 1856)	32	Three specimens collected with rotenone.	1-15

SPECIES	SITE RECORDS	ABUNDANCE	DEPTH (m)
Variola albimarginata Baissac, 1953	10, 13, 18, 25, 28, 30, 31, 36, 37	Occasional and always in low numbers.	12-90
V. louti (Forsskål, 1775)*	7, 13, 16, 25	Occasional and always in low numbers.	4-150
PSEUDOCHROMIDAE			
Amsichthys knighti (Allen, 1987)	20, 44	Collected with rotenone.	5-25
Cypho purpurescens (De Vis, 1884)	22, 30, 32	Three collected with rotenone. Photographed.	8-30
Labracinus cyclophthalmus (Müller & Troschel, 1849)*	1, 6, 11-13, 16, 17, 41	Occasional.	3-20
Lubbockichthys multisquamatus (Allen, 1987)	20, 41	Collected with rotenone.	12-30
P. bitaeniatus (Fowler, 1931)	1, 2a, 11, 12, 15, 18, 22, 28, 30, 32, 33, 37, 41, 43, 44	Occasional.	5-20
P. cyanotaenia Bleeker, 1857	8	Rare, one pair seen.	2-10
P. elongatus Lubbock, 1980	2a, 12, 22, 28, 43	Occasional.	5-25
P. fuscus (Müller and Troschel, 1849)*	2a, 3, 5, 6, 8, 9, 11-13, 15-17, 24, 25, 27, 29, 38, 40, 42, 44	Occasional, around small coral and rock outcrops.	1-30
P. marshallensis (Schultz, 1953)	2a, 12, 17, 22, 32	Occasional under rocky overhangs.	2-25
P. perspicillatus Günther, 1862	3, 9, 10, 17, 24-26, 37, 43	Occasional around rock outcrops in sand-rubble areas.	3-20
P. porphyreus Lubbock & Goldmann, 1974	1-3, 6, 7, 10, 12, 13, 15, 16, 18, 20, 22, 25, 28, 31-33, 36, 39, 43, 44	Common at base of slopes.	15-40
P. splendens (Fowler, 1931)	1, 2a, 7, 11-13, 15-18, 20-22, 25, 26, 37, 39, 41, 43	Common around coral formations.	5-30
P. tapienosoma Bleeker, 1853	8, 32	Only two seen, but has cryptic habits. One collected.	2-60
Pseudoplesiops annae (Weber, 1913)	20	Four collected with rotenone.	4-25
P. typus Bleeker, 1858	1998-99		5-30
PLESIOPIDAE			
Plesiops coeruleolineatus Rüppell, 1835*	1998-99		0-3
Plesiops corallicola Bleeker, 1853*	1998-99		0-3

SPECIES	SITE RECORDS	ABUNDANCE	DEPTH (m)
ACANTHOCLINIDAE			
Belonepterygium fasciolatum (Ogilby, 1889)	28	One specimen collected with rotenone.	5-20
CIRRHITIDAE			
Cirrhitichthys aprinus (Cuvier, 1829)	15, 16, 18, 21, 22, 28, 30, 42, 43	Occasional, usually on sponges.	5-40
C. falco Randall, 1963	7, 9-11, 15, 16, 30, 32-34	Occasional.	4-45
C. oxycephalus (Bleeker, 1855)	2a, 14, 21, 22, 25, 30, 34, 39, 44	Occasional.	2-40
Cirrhitus pinnulatus (Schneider, 1801)	1	Rare, one seen in surge zone.	0-10
Cyprinocirrhites polyactis (Bleeker, 1875)	1998-99		10-132
Oxycirrhitus typus Bleeker, 1857	2a	Rare, only one seen.	10-100
Paracirrhites arcatus (Cuvier, 1829)	10, 14, 36, 37	Occasional.	1-35
P. forsteri (Schneider, 1801)*	1-3, 7, 10-12, 14, 17, 18, 20-22, 24-26, 28, 30-34, 36, 37, 39	Moderately common, the most abundant hawkfish, but always seen in low numbers.	1-35
OPISTOGNATHIDAE			
Opistognathus sp. 1	7, 10, 12, 44	Occasional.	5-20
O. sp. 2	3	Rare, only one seen.	5-20
O. sp. 3	1998-99		5-20
O. sp. 4	1998-99		5-20
TERAPONTIDAE			
Terapon jarbua (Forsskål, 1775)*	13	About 20 seen along shore at Kri Island.	0-5
Terapon theraps Cuvier, 1829	34	One taken from floating *Sargassum*.	0-5
PRIACANTHIDAE			
Priacanthus hamrur (Forsskål, 1775)	1998-99		5-80
APOGONIDAE			
Apogon angustatus (Smith and Radcliffe, 1911)	34	Rare, only one seen.	5-30
A. aureus (Lacepède, 1802)*	2a, 14, 25, 26, 28, 30, 31, 39, 41	Occasional.	10-30

SPECIES	SITE RECORDS	ABUNDANCE	DEPTH (m)
A. bandanensis Bleeker, 1854*	28	Only a few seen at one site, but nocturnal.	3-10
A. cavitiensis (Jordan & Seale, 1907)	8	Rare, several seen in silty conditions.	8-35
A. ceramensis Bleeker, 1852*	12, 41	Only two schools seen, but shelters among mangrove roots.	0-3
A. chrysopomus Bleeker, 1854	29	Rare, only 5 seen.	2-12
A. chrysotaenia Bleeker, 1851	2a, 9, 10, 13, 14, 18, 21, 22, 24-26, 28, 30, 37	Moderately common.	1-14
A. compressus (Smith and Radcliffe, 1911)	2a, 2b, 6, 9, 15, 23, 26, 29, 40, 42-44	Moderately common.	2-20
A. crassiceps Garman, 1903	20, 29, 32, 42	Collected with rotenone.	1-30
A. cyanosoma Bleeker, 1853	2a, 6, 12, 14-17, 25, 26, 28, 30, 32, 37, 39, 43, 44	Moderately common. Phtographed.	3-15
A. dispar Fraser and Randall, 1976	20	One aggregation seen in 20 m.	12-50
A. doryssa Jordan & Seale, 1906	42	One specimen collected with rotenone.	2-25
A. exostigma Jordan and Starks, 1906	12, 22, 26, 41	Rare, 6 fish seen at four sites.	3-25
A. fleurieu (Lacepède, 1802)	2a, 37	Rare, only 2 small aggregations seen.	5-30
A. fraenatus Valenciennes, 1832	6, 22, 30, 37, 39, 44	Seen at relatively few sites, but locally common under ledges and in coral crevices.	3-35
A. fragilis Smith, 1961	19, 27, 31, 35	Rarely seen, but locally abundant. .	1-15
A. fuscus Quoy and Gaimard, 1824*	2a, 32	Rarely seen during day, but probably common at night.	3-15
A. hartzfeldi Bleeker, 1852	1998-99		1-10
A. hoeveni Bleeker, 1854	8	Rare.	1-25
A. kallopterus Bleeker, 1856	2b, 16, 26, 28, 39	Occasional, but nocturnal.	3-35
A. leptacanthus Bleeker, 1856	31, 44	Rarely encountered, but locally common among branching corals	1-12
A. melanoproctus Fraser and Randall, 1976	20	One group seen in cave at 20 m depth.	15-40
A. sp. 3	29, 30, 37, 43	Occasional. Photographed.	3-35
A. multilineatus Bleeker, 1865	3, 6, 12	Rare, but nocturnal habits.	1-5
A. nanus Allen, Kuiter, and Randall, 1994	4, 8, 19, 27	Occasional aggregations.	5-20

SPECIES	SITE RECORDS	ABUNDANCE	DEPTH (m)
A. neotes Allen, Kuiter, and Randall, 1994	8, 38, 39	Occasional aggregations.	10-25
A. nigrofasciatus Schultz, 1953	1, 2a, 3, 6, 7, 10, 11, 14, 15, 20-22, 25, 26, 28, 30-33, 36, 39, 41-44	Moderately common, one of most abundant cardinalfishes, but always in small numbers under ledges and among crevices.	2-35
A. notatus (Houttuyn, 1782)	25, 43	About 20 fish seen.	2-30
A. novemfasciatus Cuvier, 1828*	13	Rare.	0.5-3
A. ocellicaudus Allen, Kuiter, and Randall, 1994	6, 12, 39, 41-44	Occasional, but locally common, especially at sites 6 and 41. Photographed.	11-55
A. parvulus (Smith & Radcliffe, 1912)	2a, 6-12, 14-17, 29, 35, 37, 41-44	Moderately common.	2-30
A. perlitus Fraser & Lachner, 1985	1998-99		2-15
A. rhodopterus Bleeker, 1852	1998-99		10-40
A. sealei Fowler, 1918	2a, 12, 13, 25, 35, 43	Occasional.	2-12
A. selas Randall and Hayashi, 1990	26	Rare, but usually in deeper water.	20-35
A. sp. 1	12, 37	About 15 fish seen.	12-15
A. sp. 2	6	Rare, only 3 seen.	45-50
A. taeniophorus Regan, 1908	1, 33	Rare, but occurs in very shallow water and is nocturnal and therefore difficult to accurately survey.	0.5-2
A. talboti Smith, 1961	42	One collected with rotenone.	10-30
A. thermalis Cuvier, 1829*	4, 5, 12, 23	Occasional.	0-10
A. timorensis Bleeker, 1854	2b		0-3
A. trimaculatus Cuvier, 1828	23, 26	Rare, but difficult to survey due to nocturnal habitats.	2-10
A. wassinki Bleeker, 1860	2b, 12	Few sightings, but locally common on silty reefs.	3-18
Apogonichthys ocellatus (Weber, 1913)	18	One collected in 30 m depth.	2-35
Archamia biguttata Lachner, 1951	7, 26	Two schools seen in caves.	5-18
A. fucata (Cantor, 1850)	2a, 6-8, 12, 14, 23, 26, 27, 29, 31, 33, 38, 40-44	Occasional, but common at several sites.	3-60

SPECIES	SITE RECORDS	ABUNDANCE	DEPTH (m)
A. macropterus (Cuvier,1828)	13, 26, 44	Occasional.	3-15
A. zosterophora (Bleeker, 1858)	2a, 2b, 12, 19, 26, 27, 29, 31, 35, 42-44	Occasional, but common at some sites, especially 44.	2-15
Cercamia eremia (Allen, 1987)	41, 42	Collected with rotenone.	5-40
Cheilodipterus alleni Gon, 1993	6, 12, 16, 26, 31	Occasional, especially in caves and crevices on steep slopes.	1-25
C. artus Smith, 1961	2b, 6, 9, 15, 27, 29, 31, 35, 42-44	Moderately common, often among branching corals.	2-20
C. macrodon Lacepède, 1801	1, 2a, 6, 10, 12, 13, 16, 21, 25, 30, 32, 39	Moderately common, but always in low numbers (except juveniles).	4-30
C. nigrotaeniatus Smith & Radcliffe, 1912	2a, 2b, 23, 26, 27, 29	Occasional on sheltered inshore reefs. Photographed.	2-25
C. quinquelineatus Cuvier, 1828	2a, 2b, 3, 6, 9, 11-13, 15, 19, 23, 25-27, 29, 31, 32, 35-38, 40-44	Common, most abundant member of genus.	1-40
C. singapurensis Bleeker, 1859*	1998-99		2-15
Fowleria aurita (Valenciennes, 1831)	1998-99		0-15
F. punctulata (Rüppell, 1832)	1998-99		2-15
Gymnapogon sp.	1998-99		1-15
G. urospilotus Lachner, 1953	32	One collected with rotenone.	1-15
Pseudamia gelatinosa Smith, 1955	1998-99		1-40
P. hayashi Randall, Lachner and Fraser, 1985	20, 41	Several collected with rotenone,	2-64
Rhabdamia cypselurus Weber, 1909	12, 14-17, 27	Occasionally observed, but sometimes in large numbers swarming around coral bommies.	2-15
R. gracilis (Bleeker, 1856)	2a, 6, 8, 10, 12, 14, 17, 25, 26, 30, 31, 37, 39, 41, 43	Moderately common, forming large aggregations around coral heads. Photographed.	5-20
Sphaeramia nematoptera (Bleeker, 1856)	2a, 2b, 19, 23, 26, 27, 44	Occasional, but locally common among sheltered corals.	1-8
S. orbicularis (Cuvier, 1828)*	4, 5, 8, 23, 27, 31, 41	Common along sheltered shores of rocky islets and in mangroves.	0-3

SPECIES	SITE RECORDS	ABUNDANCE	DEPTH (m)
SILLAGINIDAE			
Sillago sihama (Forsskål, 1775)	1998-99		0-15
MALACANTHIDAE			
Hoplolatilus cuniculus Randall & Dooley, 1974	24, 25	Rare, several seen below 40 m.	25-115
H. purpureus Burgess, 1978	25	Rare, several pairs seen below 40 m.	18-80
H. starcki Randall and Dooley, 1974	1998-99		20-105
Malacanthus brevirostris Guichenot, 1848	7, 16, 30, 37, 39	Occasional.	10-45
M. latovittatus (Lacepède, 1798)	7, 11, 14, 20, 25, 30, 32, 33	Occasional.	5-30
ECHENEIDAE			
Echeneis naucrates Linnaeus, 1758*	14	Rare, only one seen.	0-30
CARANGIDAE			
Carangoides bajad (Forsskål, 1775)	4-7, 10-14, 16-18, 20, 26, 39, 41, 43, 44	Occasional, usually in low numbers.	5-30
C. ferdau (Forsskål, 1775)	37	Rare, only one seen.	2-40
C. fulvoguttatus (Forsskål, 1775)	2a, 4, 6, 12, 18, 20, 25-28, 31	Occasional.	5-100
C. plagiotaenia Bleeker, 1857	2a, 3, 6, 7, 12, 15, 16, 18, 20, 21, 25, 31, 37, 39, 42	Moderately common, but usually solitary fish encountered.	5-200
Caranx ignobilis (Forsskål, 1775)*	14, 28	Rare, two large adults seen.	2-80
C. melampygus Cuvier, 1833*	2a, 14, 16, 20, 33, 36, 37, 44	Occasional, usually seen solitary or in small schools. Waigeo is type locality.	1-190
C. papuensis Alleyne & Macleay, 1877	4, 8, 11, 13, 23, 29, 31	Occasional.	1-50
C. sexfasciatus Quoy and Gaimard, 1825*	1998-99	Waigeo is type locality.	3-96
Elegatis bipinnulatus (Quoy and Gaimard, 1825)	33-35	Occasional.	5-150
Gnathanodon speciosus (Forsskål, 1775)*	35	Rare, only one juvenile seen.	1-30
Scomberoides lysan (Forsskål, 1775)	15, 20, 26, 33, 37	Occasional.	1-100
Selar boops (Cuvier, 1833)	4	One large aggregation seen.	1-30
S. crumenophthalmus (Bloch, 1793)*	13, 27	Two schools encountered,	1-170

SPECIES	SITE RECORDS	ABUNDANCE	DEPTH (m)
Selaroides leptolepis (Kuhl and van Hasselt, 1833)	23	One school.	1-25
Trachinotus blochii (Lacepède, 1801)	38	Rare, 2 fish seen.	3-40
LUTJANIDAE			
Aprion virescens Valenciennes, 1830	10, 16, 37, 39	Rare, only 4 seen.	3-40
Lutjanus argentimaculatus (Forsskål, 1775)*	4, 12	Rare, but locally common at 2 sites.	1-100
L. biguttatus (Valenciennes, 1830)	2a, 2b, 4, 6, 11, 15, 23, 29, 35, 41-44	Occasional, mainly on sheltered reefs with rich corals. Especially abundant at site 43.	3-40
L. bohar (Forsskål, 1775)	1, 2a, 2b, 9-11, 13-18, 20-22, 25, 26, 28, 30-34, 36, 37, 39, 43	Common, especially abundant at site 14.	4-180
L. bouton (Lacepède, 1802)*	1998-99		5-25
L. carponotatus (Richardson, 1842)	2b, 3-6, 8-12, 15-18, 21-23, 27, 29, 31, 35, 37, 39-42	Moderately common, usually on sheltered coastal reefs.	2-35
L. decussatus (Cuvier, 1828)	1, 2a, 3, 5-16, 18-31, 36, 37, 39-44	Common, but always seen in small numbers.	3-25
L. ehrenburgi (Peters, 1869)	4, 5, 8, 9, 33	Occasional.	1-20
L. fulviflamma (Forsskål, 1775)*	9, 12, 13, 18, 28	Occasional, but locally common at a few sites.	1-35
L. fulvus (Schneider, 1801)*	1, 4, 5, 11, 13, 19, 21, 28, 30-34, 37	Moderately commmom, but usually in small numbers.	2-40
L. gibbus (Forsskål, 1775)*	1, 2a, 3, 5, 6, 8-11, 13, 14, 17, 20, 21, 25, 28, 30-34, 36, 37, 39	Moderately common.	6-40
L. johnii (Bloch, 1792)	1998-99		
L. kasmira (Forsskål, 1775)*	2a, 7, 25, 39	Occasional, ususally in low numbers.	3-265
L. lemniscatus (Valenciennes, 1828)	1998-99		1-40
L. lutjanus Bloch, 1790	33	Rare, only one seen.	5-90
L. monostigma (Cuvier, 1828)*	1, 2a, 3, 4, 7, 10, 14, 16, 18-21, 26, 28, 39-33, 36, 37, 39, 41, 43, 44	Moderately common.	5-60

SPECIES	SITE RECORDS	ABUNDANCE	DEPTH (m)
L. quinquelineatus (Bloch, 1790)	8, 13, 14, 18, 25, 39	Occasional.	5-30
L. rivulatus (Cuvier, 1828)*	4, 5, 15, 37, 39	Occasional large adults seen.	2-100
L. russelli (Bleeker, 1849)	3, 4, 8, 9, 39	Occasional solitary fish sighted.	1-80
L. semicinctus Quoy and Gaimard, 1824	2b, 5, 13, 16-18, 20, 22, 24, 26, 28, 30-34, 36, 37, 39, 41, 42, 44	Common. Waigeo is type locality.	10-40
L. vitta (Quoy and Gaimard, 1824)*	29, 38	Rare, only 2 seen. Waigeo is type locality.	8-40
Macolor macularis Fowler, 1931	1, 2a, 3, 7, 10, 13-18, 20-21, 25, 26, 28, 30-34, 36-39, 41	Common.	3-50
M. niger (Forsskål, 1775)*	10, 14, 17, 18, 36, 44	Occasional.	3-90
Paracaesio sordidus Abe & Shinohara, 1962	36	One aggregation containing about 30 fish seen in 40 m.	5-100
Pinjalo lewisi Randall, Allen, & Anderson, 1987	25	One aggregation containing about 20 fish seen in 20 m.	8-40
Symphorichthys spilurus (Günther, 1874)*	1998-99		5-60
Symphorus nematophorus (Bleeker, 1860)	3, 11, 30	Rare, a few adults seen.	5-50
CAESIONIDAE			
Caesio caerulaurea Lacepède, 1802*	2a, 2b, 3-7, 9-14, 17, 18, 20-22, 26-28, 30, 34, 36, 37, 40, 41, 43	Abundant in variety of habitats.	1-30
C. cuning (Bloch, 1791)	2a, 2b, 4-7, 9, 11-19, 21-23, 25-35, 37-44	Abundant in variety of habitats, particularly coastal reefs.	1-30
C. lunaris Cuvier, 1830	2a, 2b, 3, 6, 7, 10, 11, 13-15, 18, 20, 21, 25, 26, 28, 30, 32, 33, 37, 39, 40, 41, 43, 44	Common.	1-35
C. teres Seale, 1906	1, 2a, 3, 7, 10, 11, 13-15, 17, 18, 20-22, 28, 30-34, 36, 37, 39, 43	Common.	1-40
Dipterygonatus balteatus (Valenciennes, 1830)	1998-99		2-20
Gymnocaesio gymnoptera (Bleeker, 1856)	18, 21	Occasionally seen with mixed school of fusiliers, mainly Pterocaesio pisang.	5-30

SPECIES	SITE RECORDS	ABUNDANCE	DEPTH (m)
Pterocaesio digramma (Bleeker, 1865)	16, 25, 30	Occasional.	1-25
P. marri Schultz, 1953	1, 2a, 7, 10, 11, 13, 14, 16-18, 22, 25, 26, 30, 31-34, 40	Common.	1-35
P. pisang (Bleeker, 1853)	1, 2a, 10, 13-16, 18, 21, 22, 26, 28, 29, 31-34, 36, 37, 39, 41, 43, 44	Common in variety of habitats.	1-35
P. tessellata Carpenter, 1987	1, 2a, 7, 10, 14, 18, 21, 25, 26, 28, 30, 37	Occasional, but locally abundant.	1-35
P. tile (Cuvier, 1830)	1, 2a, 7, 10, 13, 14, 21, 22, 25, 26, 28, 30-32, 34, 37, 39, 43	Common.	1-60
GERREIDAE			
Gerres filamentosus Cuvier, 1829*	1998-99		0-10
G. oyena (Forsskål, 1775)*	6, 11, 13	Occasional in sandy areas.	0-10
HAEMULIDAE			
Diagramma pictum (Thünberg, 1792)*	5, 9, 17, 40	Occasional, in silty areas.	2-40
Plectorhinchus chaetodontoides (Lacepède, 1800)	2a, 10, 14, 16, 17, 20, 38, 39, 41, 44	Occasional.	1-40
P. chrysotaenia (Bleeker, 1855)	2a, 3, 14, 16, 21, 30, 34, 37, 39	Occasional.	4-30
P. gibbosus (Lacepède, 1802)*	1998-99		2-30
P. lessoni (Cuvier, 1830)*	14, 17, 28, 30, 32, 34, 36, 37, 39	Occasional. Waigeo is type locality.	5-35
P. lineattus (Linnaeus, 1758)	1, 3, 7, 10, 11, 14, 16, 17, 21, 25, 30, 34, 36-40	Moderately common.	2-40
P. obscurus (Günther, 1871)	34, 38	Rare, two large adults seen.	5-50
P. polytaenia (Bleeker, 1852)*	1, 2a, 3, 5, 7-11, 14-18, 21, 25, 28-30, 32, 33, 36, 39	Moderately common.	5-40
P. unicolor (Macleay, 1883)	33	Rare, about 10 adults seen in 4 m.	2-25

SPECIES	SITE RECORDS	ABUNDANCE	DEPTH (m)
P. vittatus (Linnaeus, 1758)	3, 13, 17, 25, 26, 31, 34, 37, 39, 44	Occasional.	3-30
LETHRINIDAE			
Gnathodentex aurolineatus Lacepède, 1802	25	Rare, one aggregation of 30 fish seen.	1-30
Gymnocranius grandoculus (Valenciennes, 1830)	2a, 37	Rare.	20-100
G. sp.	1998-99		15-40
Lethrinus atkinsoni Seale, 1909	13, 17, 26, 28, 31, 41	Occasional.	2-30
L. erythracanthus Valenciennes, 1830	15, 20, 25, 31, 44	Occasional.	15-120
L. erythropterus Valenciennes, 1830*	1-3, 5-7, 16, 18, 20-22, 23-28, 31, 35, 37, 40-42, 44	Moderately common.	2-30
L. harak (Forsskål, 1775)*	2a, 5, 6, 11-13, 16, 25, 26, 33, 34, 41	Moderately common on sheltered reefs near shore.	1-20
L. laticaudis Alleyne & Macleay, 1777	14	Rare, only one seen.	3-40
L. lenjan (Lacepède, 1802)	16	Rare, 3 individuals seen.	10-50
L. obsoletus (Forsskål, 1775)	5-7, 14, 17, 24, 25, 31, 39, 41	Occasional, and always in low numbers.	1-25
L. olivaceous Valenciennes, 1830*	2a, 9, 10	Occasional.	4-185
L. ornatus Valenciennes, 1830	3, 25, 39	Occasional	3-20
L. semicinctus Valenciennes, 1830*	25, 31, 33, 38, 39	Occasional, always below 30 m depth. Waigeo is type locality.	10-40
L. variegatus Valenciennes, 1830	9	Rare, but seagrass is main habitat.	1-10
L. sp. 2 (Carpenter & Allen, 1989)	30	Rare, only one seen in 30 m depth.	10-40
L. xanthocheilus Klunzinger, 1870	20, 25, 36	Rare, less than 10 fish seen.	2-25
Monotaxis grandoculis (Forsskål, 1775)	1, 2a, 2b, 3, 6, 7, 10-14, 16-18, 20-22, 24, 26-28, 30-34, 36-44	Common. The most abundant lethrinid at the Raja Ampats.	1-100
NEMIPTERIDAE			
Pentapodus emeryii (Richardson, 1843)*	8, 14, 24, 25, 28, 29, 31, 32, 34, 35, 39	Occasional.	5-40

SPECIES	SITE RECORDS	ABUNDANCE	DEPTH (m)
P. sp. (Russell, 1990)	2a, 3, 7, 10-12, 15, 16, 22, 24-26, 29, 32, 35, 38, 42	Moderately common at base of slopes over sand-rubble bottoms.	3-25
P. trivittatus (Bloch, 1791)	2a, 2b, 3-6, 8, 9, 11-13, 16, 23, 26, 27, 29, 38, 40-43	Moderately common, usually on sheltered coastal reefs.	1-15
Scolopsis affinis Peters, 1876	3, 6, 7, 9, 12, 13, 15, 16, 24-26, 30, 33, 38, 39, 42, 43	Occasional, but locally common in sandy areas.	3-60
S. bilineatus (Bloch, 1793)*	1-7, 9-11, 13-22, 24-26, 28-34, 42-44	Common.	2-20
S. ciliatus (Lacepède, 1802)	2b, 3, 4, 8, 0, 23, 40	Moderately common at sites subjected to silting.	1-30
S. lineatus Quoy and Gaimard, 1824*	2a, 2b, 6, 7, 9, 11, 13, 33, 41	Occasional on shallow reefs. Waigeio is type locality.	0-10
S. margaritifer (Cuvier, 1830)*	1, 2b, 3-19, 23-27, 29, 35, 37, 38, 40-44	Common, especially on sheltered coastal reefs. Waigeo is type locality.	2-20
S. monogramma (Kuhl and Van Hasselt, 1830)	1998-99		5-50
S. temporalis (Cuvier, 1830)*	3, 5, 8, 12, 13, 21, 29, 38	Occasional, over sand bottoms. Waigeo is type locality.	5-30
S. trilineatus Kner, 1868	1998-99		1-10
S. vosmeri (Bloch, 1792)	16	Rare, only one seen.	3-35
S. xenochrous (Günther, 1872)*	2a, 8, 9, 16, 24, 25, 28, 29, 31, 33	Occasional, usually below 20 m.	5-50
MULLIDAE			
Mulloidichthys flavolineatus (Lacepède, 1802)*	5, 11, 26, 33, 37	Occasional, usually seen in small groups.	1-40
Parupeneus barberinoides (Lacepède, 1801)	11	Rare, only one seen, but more common in seagrass habitat.	1-20
P. barberinus (Lacepède, 1801)*	1-3, 6-26, 28-33, 35-44	Common.	1-100
P. bifasciatus (Lacepède, 1801)	1-3, 6, 7, 10, 11, 13-16, 18-20, 22, 24-26, 30-44	Common.	1-80
P. cyclostomus (Lacepède, 1802)	1-3, 7, 10, 11, 13-18, 20, 25, 26, 28, 30-33, 36, 37, 39-41	Moderately common, but in lower numbers than previous two species.	2-92
P. heptacanthus (Lacepède, 1801)	3, 12, 25	Rare, less than 10 seen.	1-60

SPECIES	SITE RECORDS	ABUNDANCE	DEPTH (m)
P. indicus (Shaw, 1903)*	44	Rare, only one seen.	0-15
P. multifasciatus Bleeker, 1873	1-3, 6, 7, 9-11, 13-18, 20-22, 24-26, 28-34, 36-44	Common.	1-140
P. pleurostigma (Bennett, 1830)	13	Rare, only one seen.	5-46
Upeneus tragula Richardson, 1846*	3, 4, 12, 15, 29, 42	Occasional, but mainly found on sand bottoms away from reefs.	1-40
PEMPHERIDAE			
Parapriacanthus ransonneti Steindachner, 1870	2a, 7, 10, 12, 14, 17, 25, 30, 31, 33, 41, 43	Moderately common, forming large aggreations in caves and crevices.	5-30
Pempheris mangula Cuvier, 1829	20, 26, 33, 34, 36, 42	Moderately common, forming large aggreations in caves and crevices.	5-30
P. vanicolensis Cuvier, 1831	5, 16, 21, 28, 32, 36	Moderately common, forming large aggreations in caves and crevices.	3-38
TOXOTIDAE			
Toxotes jaculatrix (Pallas, 1767)*	5, 23, 27	Occasional where reef and mangroves in close proximity.	0-2
KYPHOSIDAE			
Kyphosus bigibbus Lacepède, 1801	16, 34, 37	Occasional.	1-20
K. cinerascens (Forsskål, 1775)	1, 2a, 11, 14, 17, 21, 30, 32-34, 37, 39, 44	Moderately common.	1-24
K. vaigiensis (Quoy and Gaimard, 1825)*	2a, 10, 13, 20, 24, 26, 28, 30, 31, 33, 34, 37	Moderately common. Waigeo is type locality.	1-20
MONODACTYLIDAE			
Monodactylus argenteus (Linnaeus, 1758)*	1998-99		0-5
CHAETODONTIDAE			
Chaetodon adiergastos Seale, 1910	1, 4-7, 12, 13, 15-18, 23, 25, 32, 33, 39, 40, 42	Moderately common.	3-25
C. auriga Forsskål, 1775	1, 2a, 3, 6-8, 12, 13, 20, 21, 33, 35, 36, 38, 40-42	Moderately common.	1-30
C. baronessa Cuvier, 1831	1-3, 5-7, 9-22, 24-26, 29, 31-44	Common, seen on most dive.	2-15

SPECIES	SITE RECORDS	ABUNDANCE	DEPTH (m)
C. bennetti Cuvier, 1831*	1, 2a, 4, 7, 11, 15, 16, 19, 20, 43	Occasional.	5-30
C. citrinellus Cuvier, 1831	1, 2a, 11, 13, 20, 24, 30-34, 39, 42	Occasional on shallow reefs affected by surge.	1-12
C. ephippium Cuvier, 1831	1-3, 5-8, 10, 13-18, 21, 22, 25, 26, 30-32, 36-38, 44	Moderately common, but never more than 2-3 pairs seen at a single site.	1-30
C. kleinii Bloch, 1790*	1-3, 5-7, 9, 10, 12-18, 20-22, 24-26, 30-39, 41-44	Commonly seen at most sites.	6-60
C. lineolatus Cuvier, 1831	4, 10, 16, 29, 35	Occasional, less common than the very similar C. oxycephalus.	2-170
C. lunula Lacepède, 1803	1, 2a, 3, 7, 9, 18, 19, 21, 22, 26, 30, 31, 36-38, 40, 41	Moderately common, but always in low numbers at each site.	1-40
C. lunulatus Quoy and Gaimard, 1824	1-3, 5-32, 35-43	Common, one of the most abundant butterflyfishes at the Raja Ampats.	1-25
C. melannotus Schneider, 1801	3-5, 9, 16-18, 20-22, 25, 26, 31, 34, 39, 41	Occasional.	2-15
C. meyeri Schneider, 1801	1, 2a, 7, 16, 22, 25, 30, 37	Occasional.	5-25
C. ocellicaudus Cuvier, 1831	2a, 2b, 4, 12, 13, 16, 19, 20, 24, 26, 31, 35, 38, 43	Moderately common.	1-15
C. octofasciatus Bloch, 1787*	2b, 4, 6, 9, 11, 12, 16, 19, 23, 26, 27, 29, 35, 38, 40-42	Moderately common at sheltered sites where reef influenced by silt.	3-20
C. ornatissimus Cuvier, 1831	2a, 3, 7, 11, 17, 20, 21, 25, 26, 30-32, 37, 39, 40	Moderaely common in rich coral areas.	1-36
C. oxycephalus Bleeker, 1853	1, 2b, 10, 15-17, 20, 24, 25, 27, 32, 33, 38, 40, 44	Moderately common, but always in low numbers.	8-30
C. punctatofasciatus Cuvier, 1831*	1, 10, 11, 14, 16, 17, 20-22, 26, 32	Occasional, usually in pairs.	6-45
C. rafflesi Bennett, 1830*	1-3, 5-7, 9-12, 14-16, 18-22, 24-26, 30-35, 40-44	Common, one of the most abundant butterflyfishes at the Raja Ampats.	1-15
C. selene Bleeker, 1853	10, 15, 22, 25, 39	Occasional, usually below 20 m.	8-50
C. semeion Bleeker, 1855	1, 2a, 2b, 11, 13, 14, 16, 22, 24, 30, 31, 37	Occasional.	1-25

SPECIES	SITE RECORDS	ABUNDANCE	DEPTH (m)
C. speculum Cuvier, 1831	1, 2a, 3-7, 9, 10, 12, 13, 16-18, 20-27, 29, 32-34, 37, 40	Moderately common.	1-30
C. trifascialis Quoy and Gaimard, 1824	1-3, 6, 7, 9-11, 13-22, 24-26, 31-34, 37, 39, 43, 44	Moderately common in areas of tabular Acropora.	2-30
C. ulietensis Cuvier, 1831	4, 9-11, 15, 16, 19-21, 23, 31, 35, 37, 39	Moderately common.	8-30
C. unimaculatus Bloch, 1787	2a, 10, 17, 26, 30, 32, 33, 37, 43	Occasional.	1-60
C. vagabundus Linnaeus, 1758*	1-3-7, 9, 10, 12-22, 24-26, 28, 30-44	Common, one of most abundant butterflyfishes at the Raja Ampats.	1-30
C. xanthurus Bleeker, 1857	7	Rare, only one seen in 25 m.	12-50
Chelmon rostratus (Linnaeus, 1758)	3-5, 15, 23, 27, 29, 35	Occasional, mainly on silty reefs of the Louisiades.	1-15
Coradion chrysozonus Cuvier, 1831	1, 2a, 3, 6-10, 12-18, 21, 22, 25, 28, 29, 31, 32, 39, 41, 42	Moderately common on sheltered reefs.	5-60
Forcipiger flavissimus Jordan and McGregor, 1898	1, 3, 7, 10, 17, 20, 25, 30-32, 34-37, 39	Occasional.	2-114
F. longirostris (Broussonet, 1782)	1, 2a, 7, 13, 17, 28, 31, 34	Occasional.	5-60
Hemitaurichthys polylepis (Bleeker, 1857)	2a, 7, 25	Rare, except about 20 seen at site 25.	3-60
Heniochus acuminatus (Linnaeus, 1758)	1, 4, 5, 9, 14, 15, 21, 22	Occasional.	2-75
H. chrysostomus Cuvier, 1831	1, 2a, 3, 5-7, 10-18, 20, 22, 23, 25-28, 30-34, 36, 39		5-40
H. diphreutes Jordan, 1903	2a, 7, 14, 28, 39, 43	Occasional, usually in aggregatons.	15-210
H. monoceros Cuvier, 1831	21, 38	Rare, only 2 pairs seen.	2-25
H. singularius Smith and Radcliffe, 1911	3, 5, 15, 21, 23, 26-28, 32, 36, 38, 42	Occasional.	12-45
H. varius (Cuvier, 1829)	1, 2a, 2b, 3, 6-22, 24-44	Common.	2-30
Parachaetodon ocellatus (Cuvier, 1831)*	8	Rare, only one seen.	5-40
POMACANTHIDAE			
Apelemichthys trimaculatus (Lacepède, 1831)	1, 2a, 7, 10, 11, 13, 14, 18, 20, 22, 24, 25, 28, 32, 34, 36, 39	Moderately common.	10-50

SPECIES	SITE RECORDS	ABUNDANCE	DEPTH (m)
Centropyge bicolor (Bloch, 1798)	1, 2a, 2b, 3, 7, 9-11, 13, 14, 16-18, 20, 22, 24-26, 28, 30-33, 34, 36, 37, 39, 44	Common.	3-35
C. bispinosus (Günther, 1860)	1, 2b, 11-13, 16, 17, 21, 22, 25, 26, 28, 31, 37	Moderately common.	10-45
C. flavicauda Fraser-Brunner, 1933	10, 28, 31, 37	Generally rare, but sometimes locally common on rubble bottoms.	10-60
C. nox (Bleeker, 1853)	1-3, 11-13, 16, 26, 29, 31, 33, 38, 40-44	Occasional.	10-70
C. tibicen (Cuvier, 1831)*	1, 2a, 2b, 7, 9-11, 13-18, 20-22, 24-26, 28, 30-34, 37, 39	Moderately common.	4-35
C. vroliki (Bleeker, 1853)	1-3, 7, 9-11, 13-18, 20-22, 24-26, 28, 30-34, 36, 37, 39, 43, 44	Common.	3-25
Chaetodontoplus dimidiatus (Bleeker, 1860)*	12, 13, 18, 25, 28, 33, 34	Occasional, usually below 25 m, but common in 10-15 m at site 34.	5-40
C. mesoleucus (Bloch, 1787)	1, 2b, 3, 4, 6-10, 12, 13, 15, 16, 18, 21-26, 29, 35, 37, 38, 40-44	Moderately common.	1-20
C. sp. (possibly just white-tailed variety of C. mesoleucus)	2a, 6, 19, 23, 37, 29, 35, 38	Occasional, usually on very sheltered bays with relatively heavy siltation.	1-20
Genicanthus lamarck Lacepède, 1798*	1, 2a, 7, 10, 13, 14, 16-18, 20, 22, 25, 26, 28, 31, 33, 39	Moderately common.	15-40
G. melanospilos (Bleeker, 1857)	12, 31	Rarely seen, but locally common at site 31.	20-50
Paracentropyge multifasciatus (Smith and Radcliffe, 1911)	1, 11, 16, 25	Occasional, but seldom noticed due to cave-dwelling habits.	10-50
Pomacanthus annularis (Bloch, 1787)	15, 23	Rare, only 3 fish seen, including 2 large adults.	1-60
Pomacanthus imperator (Bloch, 1787)*	2a, 7, 10, 14-18, 20-22, 24-26, 28, 30-34, 36-39, 41, 43, 44	Moderately common, but always in low numbers.	3-70
P. navarchus Cuvier, 1831*	1, 2a, 6, 7, 10, 11, 13-18, 20-22, 24-26, 30-32, 37, 42-44	Moderately common.	3-30
P. semicirculatus Cuvier, 1831*	3, 6, 13, 29, 32, 36	Occasional. Waigeo is type locality.	5-40

SPECIES	SITE RECORDS	ABUNDANCE	DEPTH (m)
P. sexstriatus Cuvier, 1831	1, 2a, 3-10, 12, 13, 15-23, 25, 28-30, 36-38, 42	Moderately common.	3-50
P. xanthometopon (Bleeker, 1853)*	1, 2a, 3, 11, 13, 14, 17, 21	Occasional.	5-30
Pygoplites diacanthus (Boddaert, 1772)*	1, 2a, 3, 6, 7, 10, 11, 13-18, 20-22, 24-26, 28, 30-32, 35-44	Common, one of the most abundant angelfishes in the Raja Ampats.	3-50
MUGILIDAE			
Crenimugil crenilabis (Forsskål, 1775)*	1	One school of about 15 fish seen.	0-4
Liza vaigiensis (Quoy and Gaimard, 1825)*	41	One school of about 30 fish seen. Waigeo is type locality.	0-3
Valamugil seheli (Forsskål, 1775)*	11, 40	Two schools with about 20 fish in each seen.	0-4
POMACENTRIDAE			
Abudefduf bengalensis (Bloch, 1787)	4, 5, 8, 23	Rare, about 10 seen.	0-8
A. lorenzi Hensley and Allen, 1977	4, 6, 8, 9, 11, 12, 16, 31, 42	Occasional, but locally common in shallow water next to shore.	0-6
A. notatus (Day, 1869)	1, 20, 21, 30	Occasional in rocky surge zone.	1-12
A. septemfasciatus (Cuvier, 1830)	11	Rare, but surge zone environment not regularly surveyed.	1-3
A. sexfasciatus Lacepède, 1802*	2a, 3, 6-8, 11-13, 16, 21, 29, 31, 38, 40-44	Moderately common.	1-15
A. sordidus (Forsskål, 1775)	1, 11, 16, 31	Occasional, but surge zone environment not regularly surveyed.	1-3
A. vaigiensis (Quoy and Gaimard, 1825)*	1, 2a, 3, 5-13, 15, 16, 18, 20-22, 24-26, 29-34, 36-38, 40-44	Generally common. Waigeo is the type locality.	1-12
Acanthochromis polyacantha (Bleeker, 1855)	2b, 5-7, 10, 11, 13, 14, 16-18, 23-27, 29, 40-44	Moderately common.	1-50
Amblyglyphidodon aureus (Cuvier, 1830)	1, 2a, 3, 6, 7, 10, 12, 14, 15-18, 20-22, 25, 26, 28, 32, 36, 37, 39, 42-44	Common on steep slopes, but always in low numbers.	10-35
A. batunai Allen, 1995	26	Rare.	2-12
A. curacao (Bloch, 1787)	1-3, 6, 7, 9-14, 16-21, 24-27, 29-33, 35, 37, 38, 40-44	Common.	1-15

SPECIES	SITE RECORDS	ABUNDANCE	DEPTH (m)
A. leucogaster (Bleeker, 1847)	1, 2a, 2b, 6, 7, 10, 11, 13, 14, 16-18, 20-22, 24-26, 28-33, 35-44	Common.	2-45
A. ternatensis (Bleeker, 1853)	2a, 2b, 4, 12, 19, 23, 26, 27, 29, 31, 35, 38	Common on silty inshore reefs.	2-12
Amblypomacentrus breviceps (Schlegel and Müller, 1839-44)	3, 4, 12, 29, 40, 42-44	Occasional, around debris and small coral outcrops situated on sloping silt bottoms.	2-35
Amphiprion chrysopterus Cuvier, 1830	1998-99		1-20
A. clarkii (Bennett, 1830)	1, 2a, 2b, 6, 7, 10-18- 20, 21, 24-26, 28, 30, 31, 33-38, 40, 41, 43, 44	One of the two most common anemonefishes at the Raja Ampats.	1-55
A. melanopus Bleeker, 1852	6, 15, 21, 22, 26, 32, 43	Occasional.	1-10
A. ocellaris (Cuvier, 1830)*	1, 2a, 2b, 3-5, 7, 9, 11-16, 18, 20-22, 24, 25, 29-32, 37-44	One of the two most common anemonefishes at the Raja Ampats.	1-15
A. perideraion Bleeker, 1855*	7, 11, 15-18, 21, 26, 28, 33, 34, 40	Occasional.	3-20
A. polymnus (Linnaeus, 1758)*	24, 30	Rare, but restricted to featureless silt or sand bottoms away from reefs.	2-30
A. sandaracinos Allen, 1972	2, 5, 6, 21, 30, 35, 40, 42	Occasional.	3-20
Cheiloprion labiatus (Day, 1877)	6, 13, 26	Rare, about 10 seen.	1-3
Chromis alpha Randall, 1988	1, 2a, 11, 13, 16, 33	Occasional on steep slopes. Photographed	18-95
C. amboinensis (Bleeker, 1873)	1-3, 6, 7, 9, 11, 13-16, 18, 20-22, 25, 26, 31-33, 37, 39-44	Common.	5-65
C. analis (Cuvier, 1830)	1, 2a, 14-16, 18, 20-22, 25, 26, 44	Moderately common on steep slopes.	10-70
C. atripectoralis Welander and Schultz, 1951	1, 2a, 3, 5, 6, 10, 13-18, 20, 22, 24-26, 32, 37, 42, 44	Common. .	2-15
C. atripes Fowler and Bean, 1928	2a, 10, 13, 17, 20-22, 25, 28, 30-33, 36, 37, 39	Moderately common on steep slopes.	10-35

SPECIES	SITE RECORDS	ABUNDANCE	DEPTH (m)
C. caudalis Randall, 1988	1, 7, 11, 14, 16, 20, 22, 25, 28, 32, 37, 39	Occasional on steep slopes.	20-50
C. cinerascens (Cuvier, 1830)	5	Rare, but locally common at site 5.	3-20
C. delta Randall, 1988	1, 2a, 7, 11-16, 20-22, 25, 33, 36, 37, 39	Moderately common, especially on steep slopes below about 15 m depth.	10-80
C. elerae Fowler and Bean, 1928	1, 2a, 12, 15, 18, 20, 21, 25, 33, 36, 39, 41, 44	Moderately common, always in caves and crevices on steep slopes.	12-70
C. lepidolepis Bleeker, 1877	1-3, 6, 7, 9-11, 13-15, 17, 18, 20-22, 24-26, 30-32, 34, 36, 37, 39, 43, 44	Common.	2-20
C. lineata Fowler and Bean, 1928	2a, 7, 10, 11, 13, 21, 22, 24, 25, 31-33, 37	Moderately common, usually in clear water with some wave action.	2-40
C. margaritifer Fowler, 1946	1, 2a, 3, 7, 9-11, 13, 14, 17, 20-22, 24-26, 28, 30-34, 36, 37, 39, 43	Common in clear water areas.	2-20
C. retrofasciata Weber, 1913	1-3, 7-11, 13-18, 20-22, 24-26, 31, 32, 37, 39, 43-44	Common at most sites.	5-65
C. scotochiloptera Fowler, 1918	1, 2a, 7, 10, 11, 14, 15, 18, 20-22, 25, 26, 28, 39, 43	Moderately common.	5-20
C. ternatensis (Bleeker, 1856)	1-3, 6, 7, 9-18, 20-22, 24-26, 28-34, 36-41, 43, 44	Abundant, often forming dense shoals on upper edge of steep slopes.	2-15
C. viridis (Cuvier, 1830)	1, 2b, 3, 5-7, 9, 11-13, 15, 17, 18, 20-22, 24-27, 29, 33, 37, 38, 42-44	Abundant in sheltered areas of rich coral, generally in clear water.	1-12
C. weberi Fowler and Bean, 1928	1, 2a, 5-7, 9-11, 13-18, 20-22, 24-26, 28, 30-34, 36, 37, 39, 41-44	Common.	3-25
C. xanthochira (Bleeker, 1851)	1, 7, 11, 22, 26, 37, 39	Occasional on outer slopes.	10-48

SPECIES	SITE RECORDS	ABUNDANCE	DEPTH (m)
C. xanthura (Bleeker, 1854)	1, 2a, 3, 6, 7, 10, 11, 13-18, 20-22, 24-26, 28, 30-33, 36, 37, 39, 42-44	Common, especially on steep slopes.	3-40
Chrysiptera biocellata (Quoy and Gaimard, 1824)*	1998-99		0-5
C. bleekeri (Fowler and Bean, 1928)	7, 9, 10, 13, 14, 17, 18, 22, 24, 25, 28, 30, 31, 42, 43	Moderately common on rubble bottoms below 15 m.	3-30
C. brownriggii (Bennett, 1828)	1, 2a, 16, 20, 21, 24, 31, 33, 36, 37	Moderately common, usually in shallow beach rock areas affected by surge.	0-2
C. cyanea (Quoy and Gaimard, 1824)	1, 6, 11, 13, 31, 41, 44	Occasional, usually in shallow well-sheltered areas with clear water.	0-10
C. hemicyanea (Weber, 1913)	2b, 4, 6, 12, 13, 19, 23, 26, 27, 29, 35, 40, 42	Moderately common in sheltered bays and lagoons. Photographed.	1-15
C. oxycephala (Bleeker, 1877)	4, 19, 23, 26, 27, 35, 38	Occasional in sheltered bays and lagoons.	1-16
C. parasema (Fowler, 1918)	35	Rare, except moderately common at site 35.	1-16
C. rex (Snyder, 1909)	30, 31, 36	Occasional, in surge areas off NW Waigeo.	1-6
C. rollandi (Whitley, 1961)	1-3, 6-9, 11-16, 18, 20-22, 24-26, 29, 31-33, 35, 37-44	Common, particularly on reef slopes affected by silt.	2-35
C. springeri Allen & Lubbock, 1976	6, 12, 26, 40, 42, 44	Occasional in sheltered bays and lagoons.	5-30
C. talboti (Allen, 1975)	1, 2a, 3, 6, 7, 10-18, 20-22, 24-26, 28, 30-33, 37, 39, 41-44	Common, except in silty areas.	6-35
C. unimaculata (Cuvier, 1830)	9, 13, 15, 31, 42	Occasional, but locally common on shallow reef flats.	0-2
Dascyllus aruanus (Linnaeus, 1758)*	2a, 2b, 3, 6, 9, 11, 13, 16, 19, 24-27, 29, 31, 35, 38, 40-42, 44	Common in sheltered waters, forming aggregations around small coral heads.	1-12
D. melanurus Bleeker, 1854	2b, 6, 8, 9, 11, 13, 19, 26. 27, 29, 31, 35, 42	Moderately common on sheltered reefs.	1-10
D. reticulatus (Richardson, 1846)	1, 2a, 3, 5-7, 10, 11, 13-18, 20-22, 24-26, 28-34, 36, 37, 39, 42-44	Common.	1-50
D. trimaculatus (Rüppell, 1928)	1-3, 5-7, 10-18, 20-22, 24-26, 28-44	Common.	1-55

SPECIES	SITE RECORDS	ABUNDANCE	DEPTH (m)
Dischistodus chrysopoecilus (Schlegel and Müller, 1839)	5, 6, 8, 9, 11-13, 29, 31, 40, 41	Occasional in sand-rubble areas near shallow seagrass beds.	1-5
D. fasciatus (Cuvier, 1830)*	8	Rare.	1-8
D. melanotus (Bleeker, 1858)	6, 11, 13, 26, 31, 35, 40-42	Occasional.	1-10
D. perspicillatus (Cuvier, 1830)	2a, 4, 6, 8, 11-13, 19, 23, 26, 27, 29, 31, 35, 38, 40, 41, 44	Moderately common in mixed coral-sand habitat near shore.	1-10
D. prosopotaenia (Bleeker, 1852)	8, 9, 11, 12, 26, 38	Occasional. Photographed	1-12
Hemiglyphidodon plagiometopon (Bleeker, 1852)*	2a, 2b, 4, 6, 8, 9, 12, 19, 23, 26, 27, 29, 31, 35, 38, 40, 42, 44	Moderately common, generally on sheltered reefs affected by silt.	1-20
Lepidozygus tapeinosoma (Bleeker, 1856)	28	Rare, one school of about 30 fish seen.	5-25
Neoglyphidodon crossi Allen, 1991	7, 11, 16, 22, 24-26, 30, 33	Occasional, but locally common. Photographed.	2-12
N. melas (Cuvier, 1830)*	1-3, 5-7, 9-11, 13, 15-18, 20-22, 24-27, 29-44	Moderately common, but in low numbers at each site. Photographed.	1-12
N. nigroris (Cuvier, 1830)	1-3, 6, 7, 9-18, 20-22, 24-26, 30-37, 39-44	Common.	2-23
N. oxyodon (Bleeker, 1857)	4, 6, 13, 29, 35, 41, 42	Occasional on sheltered reef flats near shore	0-4
N. thoracotaeniatus (Fowler and Bean, 1928)	2a, 2b, 11, 16, 26, 31, 44	Occasional. Photographed.	15-45
Neopomacentrus azysron (Bleeker, 1877)	1, 11, 20, 21, 24, 26, 30-33, 36	Occasional, but locally common at some sites.	1-12
N. bankieri (Richardson, 1846)	5, 8, 9	Occasional, but locally common at some sites.	3-12
N. cyanomos (Bleeker, 1856)	2a, 3, 5, 6, 8, 12, 14-18, 21, 22, 26, 29-31, 33, 39, 41-43	Moderately common.	5-18
N. filamentosus (Macleay, 1833)	4, 5, 8, 9, 23, 35, 38	Occasional, but locally common on sheltered reefs.	5-15
N. nemurus (Bleeker, 1857)	5, 8, 9, 12, 19, 21, 23, 27, 40, 42	Occasional, but locally common on sheltered inshore reefs.	1-10
N. taeniurus (Bleeker, 1856)	1998-99		0-4
Plectroglyphidodon dickii (Liénard, 1839)	1, 2a, 3, 6, 7, 10, 14, 20-22, 25, 26, 33, 34, 36, 37, 39, 43	Moderately common in rich coral areas.	1-12
P. lacrymatus (Quoy and Gaimard, 1824)	1-3, 6-11, 13-18, 20-22, 24-26, 30-37, 39, 40, 42-44	Common.	2-12

SPECIES	SITE RECORDS	ABUNDANCE	DEPTH (m)
P. leucozonus (Bleeker, 1859)	1, 20, 24, 30, 31, 36	Occasional in surge areas.	0-2
Pomacentrus adelus Allen, 1991	1-3, 5, 6, 8, 9, 11-13, 15-17, 24-27, 29-35, 37, 41-44	Common.	0-5
P. amboinensis Bleeker, 1868	1-3, 5-7, 9-22, 24-26, 28, 29, 31-35, 37-44	Abundant on sand-rubble bottoms.	2-40
P. auriventris Allen, 1991	1, 2a, 5-7, 9, 10, 13-18, 20-22, 24-26, 28, 29, 30-34, 36-39, 42, 43	Common.	0-8
P. bankanensis Bleeker, 1853	1-3, 6, 7, 9-11, 13, 14, 16, 18, 20-22, 24-26, 28, 30-34, 36, 37, 39, 43, 44	Common.	0-12
P. brachialis Cuvier, 1830	1-3, 5-7, 9-18, 20-22, 24-26, 28-34, 36, 37, 39, 43, 44	Abundant, especially in areas exposed to curents.	6-40
P. burroughi Fowler, 1918	2b, 4, 6, 9, 11, 19, 23, 26, 27, 29, 35, 38, 40, 42, 44	Moderately common, usually on silty inshore reefs.	2-16
P. chrysurus Cuvier, 1830	2b, 11, 27, 42	Occasional, around small coral or rock formations surrounded by sand.	0-3
P. coelestis Jordan and Starks, 1901	1-3, 5-7, 9-12, 15-18, 20-22, 24-26, 28, 30-35, 37, 39, 42-44	Common.	1-12
P. cuneatus Allen, 1991	4, 8, 9, 29	Occasional on silty reefs.	15-25
P. grammorhynchus Fowler, 1918	2b, 6, 13, 29, 31, 40	Occasional.	2-12
P. lepidogenys Fowler and Bean, 1928	1-3, 6, 7, 10, 11, 13, 15, 16, 18, 20-22, 24-26, 30-33, 37, 39, 41-44	Common.	1-12
P. littoralis Cuvier, 1830	4, 5, 8, 31, 41	Occasional on silty, well sheltered reefs.	0-5
P. moluccensis Bleeker, 1853	1-3, 5-7, 9-18, 20-22, 24-26, 29-35, 37, 39-44	Abundant.	1-14
P. nagasakiensis Tanaka, 1917	1-3, 6, 7, 9-18, 21, 22, 25, 26, 30-33, 36, 37, 39, 42, 44	Occasional, around isolated rocky outcrops surrounded by sand.	5-30

SPECIES	SITE RECORDS	ABUNDANCE	DEPTH (m)
P. nigromanus Weber, 1913	1-3, 5, 6, 8, 9, 11-16, 18-20, 22, 23, 26, 27, 29, 35, 37-44	Common, usually on slopes in a variety of habitats.	6-60
P. nigromarginatus Allen, 1973	1, 2a, 7, 13, 16, 20, 25, 27, 31, 35, 42	Moderately common on steep slopes.	20-50
P. opisthostigma Fowler, 1918	2b, 12, 23, 27, 42	Occasional, but locally common in sheltered, silty habitats.	10-25
P. pavo (Bloch, 1878)	2a, 8, 9, 12, 31, 35, 38, 44	Occasional, usually around coral patches in sandy lagoons.	1-16
P. philippinus Evermann and Seale, 1907	18, 30, 32	Rare, less than 10 seen.	1-12
P. reidi Fowler and Bean, 1928	1, 2a, 3, 7, 10, 11, 13-16, 18, 20, 22, 25, 28, 31-33, 36, 37, 39, 41, 42	Moderately common, usually on seaward slopes.	12-70
P. simsiang Bleeker, 1856	4-6, 8, 9, 13, 14, 26, 27, 29, 31, 40, 41	Moderately common, usually in sheltered, silty bays.	0-10
P. smithi Fowler and Bean, 1928	2b, 6, 9, 11, 12, 26, 29, 42, 44	Occasional, but locally common on sheltered reefs. Photographed.	2-14
P. taeniometopon Bleeker, 1852	23	Rare, only one seen, but frequents mangroves.	0-3
P. tripunctatus Cuvier, 1830*	4, 6, 8, 13, 27, 41	Occasional in very shallow water next to shore.	0-3
P. vaiuli Jordan and Seale, 1906	1, 7, 10, 13, 20, 22, 25, 30-34, 36, 37, 39, 42, 44	Moderately common, usually on steep outer slopes.	3-45
Premnas biaculeatus (Bloch, 1790)*	2b, 4, 9, 11, 12, 16, 19, 23, 25, 40, 43	Occasional. Photographed	1-6
Stegastes albifasciatus (Schlegel and Müller, 1839)	1998-99		0-2
S. fasciolatus (Ogilby, 1889)	1, 2a, 30, 31, 33, 36	Occasional in surge areas.	0-5
S. lividus (Bloch and Schneider, 1801)	13, 29, 40	Occasional, but locally common.	1-5
S. nigricans (Lacepède, 1802)	13, 26, 35, 40	Occasional, but locally common.	1-12
S obreptus (Whitley, 1948)	1998-99		2-6
LABRIDAE			
Anampses caeruleopunctatus Rüppell, 1828*	1, 2a, 20, 24, 36	Occasional, usually female sighted.	2-30
A. geographicus Valenciennes, 1840*	16, 17, 21, 33, 34	Rare, only 5 seen.	5-25

SPECIES	SITE RECORDS	ABUNDANCE	DEPTH (m)
A. melanurus Bleeker, 1857	20, 30, 33, 36	Rare, only 6 seen.	12-40
A. meleagrides Valenciennes, 1840	1, 2a, 10, 14, 18, 25, 26, 28, 33, 39, 43	Occasional, always in small numbers.	4-60
A. neoguinaicus Bleeker, 1878	1998-99		8-30
A. twistii Bleeker, 1856	1	Rare, only one seen.	2-30
Bodianus anthioides (Bennett, 1831)	1998-99		6-60
B. axillaris (Bennett, 1831)	1998-99		2-40
B. bilunulatus Lacepède, 1801)*	1998-99		8-108
B.bimaculatus Allen, 1973	1998-99		30-60
B. diana (Lacepède, 1802)	1, 2a, 3, 6, 7, 10, 13-18, 20-22, 24-26, 28-34, 36-39, 42-44	Common.	6-25
B. mesothorax Schneider, 1801	1, 2a, 3, 6-11, 13-17, 20-22, 24-26, 29-38, 40-44	Common.	5-30
Cheilinus chlorurus (Bloch, 1791)*	5, 13, 26, 41	Occasional.	2-30
C. fasciatus (Bloch, 1791)*	1-19, 21-32, 35, 37, 39, 41-44	Common, several adults seen on most dives.	4-40
C. oxycephalus Bleeker, 1853	9-11, 13, 15-18, 21-26, 30-32, 35, 37, 39, 43, 44	Moderately common.	1-20
C. trilobatus Lacepède, 1802	1, 2a, 2b, 6, 10, 13-16, 18, 20, 22, 24-26, 29-31, 33-36, 39	Moderately common, several adults seen on most dives.	1-20
C. undulatus Rüppell, 1835	2a, 9, 14, 21, 23, 36, 39	Rare, only 7 seen.	2-60
Cheilio inermis Forsskål, 1775*	11, 13, 33, 41	Occasional, but mostly in weed habitats.	0-3
Choerodon anchorago (Bloch, 1791)	2a, 2b, 3-9, 11-13, 15, 16, 19, 26, 27, 29, 35, 38, 40, 42-44	Moderately common, usually in silty areas.	1-25
C. schoenleinii (Valenciennes, 1839)			10-80
C. zosterophorus (Bleeker, 1868)*	10, 13, 16, 18, 24-26, 28, 29, 37-39	Occasional in small groups over sand bottoms.	5-50
Cirrhilabrus condei Allen and Randall, 1996	2b	Rare, only one seen. Photographed.	25-45

SPECIES	SITE RECORDS	ABUNDANCE	DEPTH (m)
C. cyanopleura (Bleeker, 1851)	2a, 2b, 3, 6, 9-18, 20-22, 24-26, 28-34, 36-40, 42-44	Abundant in variety of habitats, but usually areas exposed to current.	2-20
C. exquisitus Smith, 1957	7, 21, 25, 37	Occasional.	6-32
C. flavidorsalis Randall & Carpenter, 1980	31	Rare, only 2 seen.	10-30
C. tonozukai Allen & Kuiter, 1999	2a, 2b, 9, 13, 14, 17, 24, 25, 28, 30, 31	Occasional.	8-40
Coris batuensis (Bleeker, 1862)	2a, 2b, 3, 7, 9-12, 14-18, 20, 24-26, 29-34, 44	Moderately common.	3-25
C. gaimardi (Quoy and Gaimard, 1824)*	1-3, 7, 9-11, 13-15, 17, 18, 20-22, 24-26, 28, 30-34, 39, 43, 44	Moderately common.	1-50
C. pictoides Randall & Kuiter, 1982	2a, 5, 12-16, 18, 22, 24, 25, 28-33	Occasional.	8-33
Cymolutes torquatus Valenciennes, 1840	1998-99		3-20
Diproctacanthus xanthurus (Bleeker, 1856)	1-3, 6-27, 29, 31, 33, 35, 37, 38, 40-44	Common, but most abundant on protected inshore reefs.	2-15
Epibulus insidiator (Pallas, 1770)	1, 2a, 2b, 4-7, 9-33, 35-44	Common.	1-40
Gomphosus varius Lacepède, 1801	1, 2a, 3, 6, 7, 10, 11, 13-18, 20-22, 24-26, 30-34, 37, 39, 41, 44	Common.	1-30
Halicheores argus (Bloch and Schneider, 1801)	1, 16, 23, 31, 38, 41	Occasional, usually in silty, protected areas with weeds.	0-3
H. biocellatus Schultz, 1960	2, 7, 22	Rare, only 3 seen.	6-35
H. chloropterus (Bloch, 1791)*	2, 4-6, 8, 9, 11-13, 15, 16, 19, 23, 26, 27, 29, 31, 35, 38, 40-42, 44	Common, usually on sheltered inshore reefs with sand and weeds.	0-10
H. chrysus Randall, 1980	1, 2a, 3, 6, 7, 10, 13-18, 20-22, 24-26, 28, 30-34, 36, 37, 39, 41-44	Common on clean sand bottoms.	7-60
H. hartzfeldi Bleeker, 1852	3, 5, 13, 16, 36, 38, 30, 33, 37, 39, 42	Occasional on sand-rubble bottoms.	10-30

SPECIES	SITE RECORDS	ABUNDANCE	DEPTH (m)
H. bortulanus (Lacepède, 1802)*	1, 2a, 3, 7, 9-11, 13-18, 20-22, 24, 26, 28, 30-34, 36, 37, 39, 42-44	Common.	1-30
H. leucurus (Walbaum, 1792)	2a, 2b, 4, 8, 9, 15, 19, 23, 27, 29, 35, 38, 40, 42, 44	Occasional, usually in silty bays. Sometimes locally common.	2-15
H. margaritaceus (Valenciennes, 1839)	1, 2a, 2b, 6, 7, 9-11, 13-18, 20-22, 24, 25, 28, 30-34, 36, 37, 39, 43, 44	Common, usually in shallow water next to shore.	0-3
H. marginatus (Rüppell, 1835)*	1, 20, 24-26, 30, 31, 33, 34	Occasional.	1-30
H. melanochir Fowler & Bean, 1928	33	Rare, only 3 seen.	4-18
H. melanurus Bleeker, 1853	1-3, 5-18, 20-22, 24-28, 31-34, 38-44	Common.	2-15
H. melasmopomus Randall, 1980	36	Rare, only one seen.	10-55
H. miniatus Kuhl and Van Hasselt, 1839	1998-99	Rare, but locally common.	0-8
H. nebulosus Valenciennes, 1839	42	Rare, less than 10 seen.	1-40
H. nigrescens Bleeker, 1862	8	Rare, several seen in silty conditions.	1-10
H. pallidus Kuiter & Randall, 1994	1998-99		5-30
H. papilionaceus (Valenciennes, 1839)*	6, 13, 40, 41	Occasional, in shallow seagrass beds.	0-4
H. podostigma (Bleeker, 1854)	20, 25, 26, 43	Rare, less than 10 seen.	2-25
H. prosopeion Bleeker, 1853	1, 2a, 3, 6, 7, 10, 11, 13-18, 20-22, 24, 25, 30-33, 36, 37, 39, 41-44	Common.	5-40
H. richmondi Fowler & Bean, 1928	21, 42	Rare, only 4 seen.	1-15
H. scapularis Bennett, 1832*	6, 10, 11, 13, 16, 18, 20, 22, 25, 26, 31, 33, 37, 38, 40, 41, 43, 44	Moderately common, always in sandy areas.	0-15
H. solorensis (Bleeker, 1853)	1, 2a, 3, 7, 11, 13-18, 20-22, 25, 26, 28, 32-34, 37, 39, 42-44	Common, except in silty bays.	2-40

SPECIES	SITE RECORDS	ABUNDANCE	DEPTH (m)
Hemigymnus fasciatus Bloch, 1792	1, 2a, 6, 7, 17, 22, 25, 26, 31, 36, 39, 40, 44	Occasional.	1-20
H. melapterus Bloch, 1791	1, 2a, 2b, 5-7, 9-22, 24-27, 29-34, 36, 37, 39, 41-44	Common, but in low numbers at each site.	2-30
Hologymnosus doliatus Lacepède, 1801	1, 2a, 7, 10, 11, 15-18, 20, 22, 24-26, 30-34, 39	Moderately common.	4-35
Labrichthys unilineatus (Guichenot, 1847)	1-3, 6, 7, 9, 11, 13, 14, 16-18, 20-22, 24-26, 30, 31, 33, 35, 37, 40, 42-44	Common, especially in rich coral areas.	1-20
Labroides bicolor Fowler and Bean, 1928	1, 2a, 2b, 16-18, 25, 26, 30-33, 36, 37, 39-44	Occasional, generally in smaller numbers than other *Labroides* species.	2-40
L. dimidiatus (Valenciennes, 1839)*	1, 2a, 3, 5-27, 29-44	Moderately common.	1-40
L. pectoralis Randall and Springer, 1975	2a, 13, 18, 20, 21, 25, 26, 30, 31, 37, 44	Occasional.	2-28
Labropsis alleni Randall, 1981	26	Rare, about 5 seen.	4-52
Leptojulis cyanopleura (Bleeker, 1853)	1, 2a, 6, 8, 9, 12, 16, 26, 31, 42, 43	Occasional, but easliy overlooked due to sandy habitat.	15-80
Macropharyngodon meleagris (Valenciennes, 1839)	1, 2a, 7, 14, 16, 17, 20, 22, 24-26, 30-34, 36, 43	Moderately common, but always in small numbers at each site.	1-30
M. negrosensis Herre, 1932	2a, 2b, 3, 5, 7, 9, 10, 13, 16-18, 24-26, 31, 34, 37, 39	Moderately common, but always in small numbers at each site.	8-30
Novaculichthys macrolepidotus (Bloch, 1791)*	1998-99		1-12
N. taeniourus (Lacepède, 1802)	2a, 2b, 7, 10, 15, 18, 20-22, 24-26, 30-33	Occasional.	1-14
Oxycheilinus bimaculatus Valenciennes, 1840	2a, 3, 8, 9, 24, 26, 30	Occasional, around rock and coral outcrops on sandy or rubble bottoms.	2-110
O. celebicus Bleeker, 1853	2a, 2b, 11, 12, 15, 16, 18, 19, 23, 26, 27, 29, 35, 38, 40-44	Moderately common on sheltered inshore reefs.	3-30

SPECIES	SITE RECORDS	ABUNDANCE	DEPTH (m)
O. diagrammus (Lacepède, 1802)	2a, 11, 14, 15, 18, 20, 22, 25, 29, 31-34	Occasional.	3-120
O. orientalis (Günther, 1862)	2a, 2b, 3, 6, 8, 9, 12, 15, 16, 22, 24, 25, 31, 33, 42	Occasional.	15-70
O. sp.	12, 15, 19, 26, 27, 42	Occasional on sand-rubble bottoms of silty bays and lagoons. Photographed.	5-35
O. unifasciatus (Streets, 1877)	18	Rare, only one seen.	3-80
Parachelinus cyaneus Kuiter & Allen, 1999	15, 24, 29, 31	Occasional.	8-40
P. filamentosus Allen, 1974	1, 2b, 6, 8, 13-15, 18, 22, 24-26, 29, 33, 37, 42, 44	Moderately common, usually in rubble areas.	10-50
Pseudocheilinops ataenia Schultz, 1960	27	Rare, only one seen.	5-25
Pseudocheilinus evanidus Jordan and Evermann, 1902	1, 2a, 7, 10, 12-14, 16, 20, 22, 24, 25, 28, 30-34, 36, 37, 39, 41, 42	Moderately common.	6-40
P. hexataenia (Bleeker, 1857)	1-3, 9-18, 20-22, 24-26, 28-34, 36-39, 41, 44	Moderately common, only a few seen on each dive, but has cryptic habits.	2-35
Pseudocoris heteroptera (Bleeker, 1857)	1998-99		10-30
P. philippina Fowler & Bean, 1928	1998-99		8-35
P. yamashiroi (Schmidt, 1930)	7, 14, 16, 18, 22, 39	Occasional, but locally common.	10-30
Pseudodax moluccanus (Valenciennes, 1840)	1, 7, 13, 17, 22, 24, 25, 30, 31, 32, 36, 39	Occasional, always in low numbers.	3-40
Pseudojuloides kaleidos Randall & Kuiter, 1994	24, 31	Rare, only 3 seen.	8-40
Pteragogus cryptus Randall, 1981	2b, 6, 29, 42	Rarely seen, but has cryptic habits.	4-65
P. enneacanthus (Bleeker, 1856)	16	Rarely seen, but has cryptic habits.	5-40
Stethojulis bandanensis (Bleeker, 1851)	3, 7, 10, 13, 20, 25, 32, 33, 36, 37, 39	Moderately common.	0-30
S. interrupta (Bleeker, 1851)	13, 16, 25	Occasional.	4-25

SPECIES	SITE RECORDS	ABUNDANCE	DEPTH (m)
S. strigiventer (Bennett, 1832)	5, 7, 9-13, 15, 16, 21, 22, 24-26, 28, 31, 33, 43, 44	Moderately common.	0-6
S. trilineata (Bloch and Schneider, 1801)*	1, 2a, 3, 6, 9-11, 13, 14, 20, 21, 24-26, 30-34, 36, 39, 40, 41	Moderately common.	1-10
Thalassoma amblycephalum (Bleeker, 1856)	1, 2a, 3, 7, 10, 12, 14, 16, 17, 20, 21, 25, 26, 28, 30-34, 36, 37, 39, 44	Common.	1-15
T. hardwicke (Bennett, 1828)*	1-3, 6-16, 18, 20-22, 24-26, 30-34, 37, 39, 40-44	Common.	0-15
T. jansenii Bleeker, 1856	1, 2a, 7, 11, 14-16, 20-22, 24-26, 30-33, 36, 39	Moderately common, usually in very shallow water exposed to surge.	0-15
T. lunare (Linnaeus, 1758)	1-3, 6-18, 20-22, 24-26, 28-39, 40, 42-44	Common.	1-30
T. purpureum (Forsskål, 1775)	21, 30, 33, 34	Occasional in surge areas.	2-20
T. quinquevittatum (Lay and Bennett, 1839)	1, 36	Rare, except locally common at 2 sites exposed to surge.	0-18
Wetmorella albofasciata Schultz & Marshall, 1954	1998-99		5-40
Xyrichtys pavo Valenciennes, 1839	33	Rare, only one seen.	2-25
SCARIDAE			
Bolbometopon muricatum (Valenciennes, 1840)	2a, 4, 5, 9, 11, 14, 15, 21, 23, 29, 30, 32, 36-38	Occasional, either lone fish or in groups of up to about 5-15 large adults.	1-30
Calotomus carolinus (Valenciennes, 1839)	13	Rare, only one seen.	4-30
Cetoscarus bicolor (Rüppell, 1828)	1-3, 9-11, 13, 14, 16, 17, 21, 26, 31, 32, 37, 39, 44	Moderately common, but usually in small numbers.	1-30
Chlorurus bleekeri (de Beaufort, 1940)	1-3, 6, 7, 10, 11, 13-20, 22, 25-27, 29, 31, 33, 35, 37, 38, 40-44	Common.	2-30
C. bowersi (Snyder, 1909)	11, 16, 24, 26, 29, 31, 35, 42, 44	Occasional.	3-20

SPECIES	SITE RECORDS	ABUNDANCE		DEPTH (m)
C. japanensis (Bloch, 1789)	14, 18, 20, 22, 25, 26, 28, 30, 31, 33, 34	Occasional.		3-15
C. microrhinos (Bleeker, 1854)	2a, 14, 16-18, 20-22, 25, 26, 30-32, 37, 39, 43, 44	Moderately common.		2-35
C. sordidus (Forsskål, 1775)	1-3, 6, 7, 9-11, 13-22, 24-26, 29-34, 36-44	Common.		1-25
Hipposcarus longiceps (Bleeker, 1862)*	2a, 2b, 3, 10, 11, 13, 28, 31, 36, 37	Moderately common at sites adjacent to sandy bottoms. Waigeo is type locality.		5-40
Leptoscarus vaigiensis (Quoy & Gaimard, 1824)	6, 13	Rare, but found in seagrass or weedy areas. Waigeo is type locality.		1-20
Scarus chameleon Choat and Randall, 1986)*	10, 17, 25, 26, 31, 39	Occasional.		3-15
S. dimidiatus Bleeker, 1859	1, 2a, 2b, 6-8, 10-16, 18-22, 24-27, 29-33, 37, 40, 41, 44	Common.		1-15
S. flavipectoralis Schultz, 1958	1-3, 5-7, 10, 11, 13-20, 22, 24-27, 29-35, 37-44	Common, one of most abundant parrotfishes at the Raja Ampats.		8-40
S. forsteni (Bleeker, 1861)	1, 2, 10, 13, 14, 18, 20, 24, 25, 30-32, 34, 36, 39, 44	Moderately common.		3-30
S. frenatus Lacepède, 1802	1, 6, 7, 9-11, 13, 14, 18, 20-22, 24, 25, 30-32, 37, 43	Moderately common.		3-25
S. ghobban Forsskål, 1775	1-23, 25, 26, 28, 29, 33, 36, 38, 40-44	Common.		3-30
S. globiceps Valenciennes, 1840	7	Rare, only one male seen.		2-15
S. hypselopterus Bleeker, 1853	16, 19, 23, 26	Occasional.		4-20
S. niger Forsskål, 1775	1, 2, 7, 10, 11, 14-22, 24-26, 30-40, 42-44	Common.		2-20
S. oviceps Valenciennes, 1839	2a, 2b, 6, 7, 10, 11, 13-16, 18, 21, 22, 24-26, 37, 41, 43, 44	Common.		1-12
S. prasiognathos Valenciennes, 1839	16, 31, 41	Rare, less than 10 seen.		4-25
S. psittacus Forsskål, 1775*	1, 2, 18, 24, 25, 28, 30, 33, 34, 39, 43	Occasional.		4-25

SPECIES	SITE RECORDS	ABUNDANCE	DEPTH (m)
S. quoyi Valenciennes, 1840	5-8, 10-12, 15, 16, 19, 20, 24-26, 29, 31, 35, 37, 38, 40, 42	Moderately common, usually on protected inshore reefs with increased turbidity.	4-18
S. rivulatus Valenciennes, 1840	5, 15, 40, 41, 44	Occasional. Photographed.	5-20
S. rubroviolaceus Bleeker, 1849	1, 2a, 7, 13, 17, 18, 20, 21, 24-26, 28, 30,34, 36, 39, 43, 44	Moderately common.	1-30
S. schlegeli (Bleeker, 1861)	2a, 6, 22, 30, 32, 39	Occasional.	1-45
S. spinus (Kner, 1868)	1, 2a, 7, 10, 15, 17, 21, 22, 24, 25, 31, 32, 37	Occasional.	2-18
S. tricolor Bleeker, 1849	7, 15	Rare, 2 adult pairs seen.	8-40
TRICHONOTIDAE			
Trichonotus elegans Shimada & Yoshino, 1984	42, 43	Locally common at 2 sites. Inconspicuous resident of sandy slopes.	5-25
Trichonotus setiger (Bloch & Schneider, 1801)	1998-99		5-25
PINGUIPEDIDAE			
Parapercis clathrata Ogilby, 1911	1, 2a, 3, 7, 10, 13, 14, 18, 20, 22, 30, 31-34, 49, 43	Occasional.	3-50
P. cylindrica (Bloch, 1792)*	13, 15, 44	Occasional in weed-sand areas.	0-20
P. hexophthalma (Cuvier, 1829)	3, 10, 11, 13, 18, 25, 44	Occasional.	5-25
P. millepunctata (Günther, 1860)	7, 20-22, 24, 25, 30, 33, 34, 36, 43, 44	Occasional.	3-50
P. schauinslandi (Steindachner, 1900)	2a, 13, 30	Rare, only 5 seen in 30-40 depth.	15-80
P. sp. 1	3, 9, 10-12, 16, 23, 26, 29, 30, 35, 37-39, 42-44	Occasional. An undescribed species ranging widely in the Indo-Australian Archipelago.	5-25
P. sp. 2	12, 29	Rare.	5-25
P. tetracantha (Lacepède, 1800)	10, 13-15, 17, 18, 24, 25, 28, 30, 37, 39, 44	Occasional.	8-40
P. xanthozona (Bleeker, 1849)	3, 5, 6, 8, 9, 12, 16	Occasional on sheltered reefs.	1-15

SPECIES	SITE RECORDS	ABUNDANCE	DEPTH (m)
PHOLIDICHTHYIDAE			
Pholidichthys leucotaenia Bleeker, 1856	12, 14, 15, 18, 21, 22, 24, 28, 31-33, 37, 39, 41	Moderately common, but usually only juveniles seen.	1-40
TRIPTERYGIIDAE			
Enneapterygius philippinus (Peters, 1869)*	1998-99	Two collected with rotenone.	8-37
E. rubricauda Shen & Wu, 1994	32		0-32
E. sp.	30	One collected with rotenone.	5-10
E. ziegleri Fricke, 1994	1998-99		0-2
Helcogramma striata Hansen, 1986	1, 2a, 10, 21, 22, 23	Occasional, but inconspicuous.	1-20
H. sp	32	One collected with rotenone.	5-15
Ucla xenogrammus Holleman, 1993	19	Rare, only one seen.	2-40
BLENNIIDAE			
Aspidontus taeniatus Quoy & Gaimard, 1834	28, 32	Rare, only 2 seen.	1-25
Atrosalarias fuscus (Rüppell, 1835)	6, 28, 41-44	Occasional in rich coral areas.	1-12
Blenniella chrysospilos (Bleeker, 1857)	22	Rare, but not readily observed due to shallow wave-swept habitat.	0-3
Cirripectes castaneus Valenciennes, 1836	20, 25, 30, 37	Occasional.	1-5
C. filamentosus (Alleyne & Macleay, 1877)	36, 42	Occasional.	1-20
C. polyzona (Bleeker, 1868)	24, 33	Occasional.	0-3
C. quagga (Fowler & Ball, 1924)	31, 32	Occasional.	1-5
C. stigmaticus Strasburg and Schultz, 1953	1, 26, 31	Occasional.	0-5
Crossosalarias macrospilus Smith-Vaniz and Springer, 1971	1, 36	Rare, only two seen.	1-25
Ecsenius bandanus Springer, 1971	6, 8, 12, 43, 44	Occasional.	2-15
E. bathi Springer, 1988	17, 18, 22, 25, 28, 39, 40	Occasional. Photographed.	3-25
E. bicolor (Day, 1838)	1, 2a, 10, 11, 14, 15, 17, 18, 21, 22, 25, 26, 30, 31, 33, 34, 39, 43	Moderately common.	3-20
E. lividinalis Chapman and Schultz, 1952	2a, 9, 10, 12, 13, 15, 17, 18, 22, 25, 29, 42-44	Moderately common in rich coral areas, usually among branches of staghorn *Acropora*. Photographed.	2-15

SPECIES	SITE RECORDS	ABUNDANCE	DEPTH (m)
E. namiyei (Jordan and Evermann, 1903)	1, 2a, 3, 6, 12, 15, 18, 21, 22, 25, 31, 33, 36, 42, 44	Moderately common.	5-30
E. stigmatura Fowler, 1952	3, 5, 6, 8, 9, 12, 15, 19, 23, 38, 40-44	Moderately common. Photographed.	2-30
E. trilineatus Springer, 1972	3, 6, 10, 11, 18, 19, 21, 32, 43	Occasional.	2-20
E. yaeyamensis (Aoyagi, 1954)	2a, 6, 21, 25, 30, 32	Occasional.	1-15
Entomacrodus striatus (Quoy and Gaimard, 1836)	1998-99		0-2
Istiblennius edentulus Bloch and Schneider, 1801*	1998-99		0-2
I. lineatus (Valenciennes, 1836)*	1998-99		0-2
Meiacanthus atrodorsalis (Günther, 1877)	3, 6, 15, 40, 42, 44	Moderately common.	1-20
M. crinitus Smith-Vaniz, 1987	12, 13, 19, 23, 26, 27, 29, 38, 40, 42	Occasional.	1-20
M. ditrema Smith-Vaniz, 1976	1998-99		5-30
M. grammistes (Valenciennes, 1836)	2a, 2b, 3, 6, 7, 9-12, 15, 16, 18, 20-22, 25, 28, 30, 31, 33, 34, 40, 42	Moderately common.	1-20
Petroscirtes breviceps (Valenciennes, 1836)	15, 34	Rare, but often found in weed habitat.	1-10
Plagiotremus rhinorhynchus (Bleeker, 1852)*	2a, 2b, 3, 7, 8, 11-13, 15, 17, 20-22, 25-34, 36, 39-44	Common, but alway in low numbers.	1-40
P. tapeinosoma (Bleeker, 1857)	1, 7, 20	Rare, only 3 seen.	1-25
Salarias fasciatus (Bloch, 1786)	2b, 6, 26, 40	Rare, only 4 seen.	0-8
S patzneri Bath, 1992	4, 5, 8, 23, 27, 35	Occasional.	0-10
S. ramosus Bath, 1992	29	Rare, only one seen. Photographed.	3-25
S. segmentatus Bath & Randall, 1991	4, 23, 38	Occasional.	0-5
S. sibogae Bath, 1992	19	Rare. Photographed.	0-5
CALLIONYMIDAE			
Anaora tentaculata Gray, 1835	1998-99		
Callionymus ennactis Bleeker, 1879*	12, 43	Two specimens collected with dipnet.	0-20

SPECIES	SITE RECORDS	ABUNDANCE	DEPTH (m)
C. pleurostictus Fricke, 1992	1998-99		0-15
Synchiropus morrisoni Schultz, 1960	1998-99		2-20
S. ocellatus (Pallas, 1770)	2a	Rare, only one seen.	0-20
S. splendidus (Herre, 1927)	1998-99		1-18
GOBIIDAE			
Acentrogobius janthinopterus (Bleeker, 1852)	1998-99		0-3
Amblyeleotris arcupinna Mohlmann & Munday, 1999	12	Rare, only one seen.	8-35
A. fasciata (Herre, 1953)	9, 24	Rare, only 2 seen.	3-20
A. fontanesii (Bleeker, 1852)	40	Locally common at site 40. Photographed.	5-30
A. guttata (Fowler, 1938)	3, 13, 15, 16, 39, 43	Occasional.	10-35
A. gymnocephala (Bleeker, 1853)	8, 43	Locally common at 2 sites.	5-30
A. latifasciata Polunin & Lubbock, 1979	2a, 9	Rare, about 5 seen.	5-30
A. periophthalma (Bleeker, 1853)	10, 12, 15, 24, 29	Occasional, locally common in some sandy areas.	8-15
A. steinitzi (Klausewitz, 1974)	12, 25, 35, 43, 44	Occasional, locally common in some sandy areas.	6-30
A. wheeleri (Polunin and Lubbock, 1977)	1, 3, 36	Rare, only 3 seen.	5-20
A. yanoi Aonuma and Yoshino, 1996	13, 14, 16	Rare, only 3 seen.	10-40
Amblygobius buanensis (Herre, 1927)	4, 8, 12, 23	Occasional.	1-5
A. bynoensis (Richardson, 1844)	12, 31, 41	Occasional.	0-5
A. decussatus (Bleeker, 1855)*	2a, 4, 9, 11, 12, 19, 26-29, 35, 38, 40-42	Moderately common in sheltered silty areas.	3-20
A. esakiae Herre, 1939	1998-99		0-10
A. nocturnus (Herre, 1945)	4, 19, 23, 27	Occasional in strongly silted areas. Photographed.	3-30
A. phalaena (Valenciennes, 1837)	2a, 2b, 5, 12, 24, 29, 31, 38, 41	Occasional.	1-20
A. rainfordi (Whitley, 1940)	2a, 2b, 3, 6, 8, 11-13, 15, 21, 23, 26, 27, 38, 40-44	Occasional, always in low numbers.	5-25
Asterropteryx bipunctatus Allen and Munday, 1996	1998-99		15-40
A. semipunctatus Rüppell, 1830	5	Generally rare, but common at one site	1-10

SPECIES	SITE RECORDS	ABUNDANCE	DEPTH (m)
A. striatus Allen and Munday, 1996	3, 6, 12, 42-44	Occasional, but locally abundant.	5-20
Bathygobius cocosensis (Bleeker, 1854)	1998-99		0-2
B. cyclopterus (Valenciennes, 1837)	1998-99		0-2
Bryaninops amplus Larson, 1985	2, 3	Seen only twice, but difficult to detect. No doubt common wherever seawhips are abundant.	10-40
B. loki Larson, 1985	1998-99		6-45
B. natans Larson, 1986	6, 42, 44	Rare, but relatively inconspicuous due to tiny size.	6-27
B. tigris Larson, 1985			5-40
B. yongei (Davis & Cohen, 1968)	8, 12	Seen only twice, but difficult to detect. No doubt common wherever seawhips are abundant.	5-40
Callogobius maculipinnis (Fowler, 1918)	41	One specimen collected with rotenone.	3-25
C. inframaculatus Randall, 1994	1998-99		5-30
Coryphopterus duospilus Hoese and Reader, 1985	28	One collected with rotenone.	5-25
C. neophytus (Günther, 1877)	2	Rare.	2-15
C. signipinnis Hoese and Obika, 1988	11, 12, 15, 16, 18, 19, 21, 23, 25, 27, 29, 38, 42-44	Moderately common.	10-30
C. maximus Randall, 2001	11, 41	Two specimens collected with rotenone.	5-25
C. melacron Randall, 2001	18, 21, 27, 43	Occasional.	4-20
Cryptocentroides insignis Seale, 1910	1998-99		3-15
Cryptocentrus cinctus (Herre, 1936)	4, 12	Rare, but sand habitat not adequately surveyed.	2-15
C. fasciatus (Playfair & Günther, 1867)	12	Rare, but sand habitat not adequately surveyed.	2-15
C. leptocephalus Bleeker, 1876*	12, 23, 41	Occasional, but sand habitat not adequately surveyed.	0-10
C. leucostictus (Günther, 1871)	3	Rare, but sand habitat not adequately surveyed.	1-15
C. octofasciatus Regan, 1908	12, 23, 41	Occasional, but sand habitat not adequately surveyed.	1-5
C. sp. 1 (spots on opercle)	4	Rare, but sand habitat not adequately surveyed.	2-10
C. sp. 2 (blue spots)	4, 23	Rare, but sand habitat not adequately surveyed. Photographed.	1-10
C. sp. 3 (yellowish)	40	Rare, only one seen. Photographed.	18-25

SPECIES	SITE RECORDS	ABUNDANCE	DEPTH (m)
C. sp. 4	40	Rare, only 6 seen. Photographed.	18-25
C. strigilliceps (Jordan and Seale, 1906)	1998-99		1-6
Ctenogobiops aurocingulus (Herre, 1935)	1998-99		2-15
C. feroculus Lubbock and Polunin, 1977	7, 12, 41	Rare, less than 10 seen.	2-15
C. pomastictus Lubbock and Polunin, 1977	12, 29	Rare, less than 10 seen.	2-20
Eviota albolineata Jewett and Lachner, 1983	3, 5, 7, 11, 16, 39, 42	Noticed on several occasions, but easily missed due to small size.	1-10
E. bifasciata Lachner and Karnella, 1980	43, 44	Rare, only 3 aggregations seen.	5-25
E. guttata Lachner and Karnella, 1978	13, 21, 26, 30, 43, 44	Noticed on only a few occasions, but easily missed due to small size.	3-15
E. nigriventris Giltay, 1933	40, 42	Rare, only 2 groups seen.	4-20
E. pellucida Larson, 1976	6, 9, 13, 16, 18, 23, 25, 27, 29, 40, 43, 44	Occasional.	3-20
E. prasina (Kluzinger, 1871)	31	Noticed once, but easily missed due to small size.	3-20
E. prasites Jordan and Seale, 1906	11, 24, 31, 40, 43	Noticed on several occasions, but easily missed due to small size.	3-15
E. sebreei Jordan and Seale, 1906	6, 10, 11, 18, 22, 26, 34, 42	Noticed on several occasions, but easily missed due to small size.	3-20
E. sp. 1	6, 19, 23, 35, 42-44	Occasional. A new species.	4-12
E. sp. 2	19, 23, 27	Noticed on 3 occasions, but easily missed due to small size.	1-12
E. sp. 3	20	Two collected with rotenone.	8-10
E. sparsa Jewett and Lachner, 1983	20	One collected with rotenone.	8-10
Exyrias bellisimus (Smith, 1959)	2b, 4, 19, 23	Occasional on silty reefs.	1-25
Exyrias puntang (Bleeker, 1851)*	1998-99		0-5
E. sp.	2b	Rare, only one seen.	3-30
Gnatholepis anjerensis Bleeker, 1851	4, 5	Rarely seen, but locally common.	1-45
G. cauerensis Bleeker, 1853	16, 43, 44	Rarely seen, but locally common.	1-45
Gobiodon okinawae Sawada, Arai and Abe, 1973	19, 26	Relatively rare, but a secretive species that is easily overlooked.	2-12
G. unicolor (Castelnau, 1873)	1998-99		2-12
Istigobius decoratus (Herre, 1927)*	19, 30	Rarely noticed, but probably more common.	1-18

SPECIES	SITE RECORDS	ABUNDANCE	DEPTH (m)
I. ornatus (Rüppell, 1830)*	8, 23	Rarely noticed, but probably more common.	0-5
I. rigilius (Herre, 1953)	10, 15, 21, 28, 32, 40, 42	Occasional.	0-30
Luposicya lupus Smith, 1959	8, 11, 19, 38	Only a few seen, but easily escapes notice due to small size. Commensal with sponges (Phyllospongia).	2-18
Macrodontogobius wilburi Herre, 1936	2b, 4, 8, 12, 19, 23, 27, 31, 35, 38, 40	Occasional in silty areas.	2-15
Mahidolia mystacina (Valenciennes, 1837)	40	Rare, but sand habitat not adequately surveyed.	1-20
Oplopomus oplopomus (Valenciennes, 1837)	31, 40	Rarely noticed, but sand habitat not adequately surveyed.	2-25
Periophthalmus argentilineatus (Valenciennes, 1837)*	31,	Recorded only once, but mainly resident of mangroves.	0-2
P. kalolo Lesson, 1830*	1998-99	Waigeio is type locality.	0-2
Phyllogobius platycephalops (Smith, 1964)	15	Only a few seen, but easily escapes notice due to small size. Commensal with sponges (Phyllospongia).	3-20
Pleurosicya labiata (Weber, 1913)	40, 42	Only a few seen, but easily escapes notice due to small size. Commensal with sponges (Ianthella).	4-30
P. elongata Larson, 1990	43	Only a few seen, but easily escapes notice due to small size. Commensal with sponges (Xestosponga).	10-40
P. mossambica Smith, 1959	43	Only a few seen, but easily escapes notice due to small size.	1-35
Priolepis fallacincta Winterbottom & Burridge, 1992	20	One collected with rotenone.	1-25
Sueviota atronasus Winterbottom & Hoese, 1988	20	One collected with rotenone.	10-25
Signigobius biocellatus Hoese and Allen, 1977	19, 29, 35, 38	Occasional on silty bottoms.	2-30
Stonogobiops xanthorhinica Hoese and Randall, 1982	42, 43	Rare, only 2 seen.	12-60
Tomiyamichthys oni (Tomiyama, 1936)	1998-99		5-30
Trimma anaima Winterbottom, 2000	42	Only one seen, but easily overlooked.	3-35
T. benjamini Winterbottom, 1996	1998-99		10-24
T. emeryi Winterbottom, 1984	20, 41, 42	Several specimens collected with rotenone.	5-25
T. griffthsi Winterbottom, 1984	6, 19, 42	Occasional, but is easily overlooked due to small size and secretive habits. Photographed.	10-40

Appendix 4

SPECIES	SITE RECORDS	ABUNDANCE	DEPTH (m)
T. halonevum Winterbottom, 2000	1998-99		1-45
T. macrophthalma (Tomiyama, 1936)	28	Collected with rotenone.	5-30
T. naudei Smith, 1957	1998-99		3-25
T. okinawae (Aoyagi, 1949)	32	Collected with rotenone.	5-30
T. rubromaculata Allen and Munday, 1995	6, 44	Generally rare, but locally common at 2 sites.	20-35
T. sp. 1	15, 20, 42	Collected with rotenone.	8-40
T. sp. 2	42	Collected with rotenone.	5-20
T. sp. 3	20, 28, 29, 41	Collected with rotenone.	5-15
T. sp. 4	20, 28, 42	Collected with rotenone.	8-25
T. striata (Herre, 1945)	2b, 12, 15, 19, 38, 40, 42,	Occasional, but easily overlooked due to small size and secretive habits.	2-25
T. taylori Lobel, 1979	20	Collected with rotenone.	15-50
T. tevegae Cohen and Davis, 1969	6, 15, 19, 20, 38, 41, 42	Occasional, but easily overlooked due to small size and secretive habits.	8-45
Valenciennea bella Hoese & Larson, 1994	16, 31	Rare, only 2 seen.	10-35
V. helsdingenii (Bleeker, 1858)*	1998-99		1-30
V. muralis (Valenciennes, 1837)	4, 5, 23, 35	Occasional in shallow sandy areas.	1-15
V. puellaris (Tomiyama, 1936)	3, 8, 9, 12, 15, 25, 26, 39, 42, 43	Occasional.	2-30
V. randalli Hoese and Larson, 1994	38	Rare, several seen in 30 m depth.	8-30
V. sexguttata (Valenciennes, 1837)	13, 31, 41, 44	Occasional.	1-10
V. strigata (Broussonet, 1782)*	1-3, 7, 9, 10, 12, 14-18, 20, 24, 25, 30-34, 38, 43, 44	Occasional, in relatively low numbers at each site.	1-25
Vanderhorstia lanceolata Yanagisawa, 1978	1998-99		4-20
MICRODESMIDAE			
Aioliops megastigma Rennis and Hoese, 1987	19, 23, 27, 38	Occasional.	1-15

SPECIES	SITE RECORDS	ABUNDANCE	DEPTH (m)
Gunnelichthys pleurotaenia Bleeker, 1858	42	Apparently rare, but inconspicuous.	2-30
PTERELEOTRIDAE			
Nemateleotris magnifica Fowler, 1938	1, 13	Rare, only 2 seen.	6-61
Parioglossus formosus (Smith, 1931)	4, 8	Locally common, but easily overlooked.	0-1
P. philippinus (Herre, 1940)	8, 19, 31, 42	Occasional.	0-1
Ptereleotris evides (Jordan and Hubbs, 1925)	1, 2, 6, 10, 15, 16, 18, 21, 22, 25, 30, 31, 36, 38, 39, 42-44	Moderately common.	2-15
P. hanae (Jordan & Snyder, 1901)	43	Rare, only one seen. Photographed.	3-43
P. heteroptera (Bleeker, 1855)	10, 18, 21, 25, 30, 37	Occasional, usually below 25 m depth.	6-50
P. microlepis Bleeker, 1856	2b, 4, 5, 8, 9, 27, 33, 35, 37	Occasional.	1-22
P. zebra (Fowler, 1938)	36, 44	Rarely seen, but locally common.	2-10
XENISTHMIDAE			
Xenisthmus polyzonatus (Klunzinger, 1871)	20, 28	Collected with rotenone.	5-20
EPHIPPIDAE			
Platax batavianus (Cuvier, 1831)	9, 10, 12	Rare, 3 large adult seen.	1-40
P. boersi Bleeker, 1852	2a, 4, 9, 12, 13, 15	Occasional.	1-20
P. orbicularis (Forsskål, 1775)*	13	Generally rare, but several small schools seen at site 13. Photographed	1-30
P. pinnatus (Linnaeus, 1758)	8, 10, 12, 15, 17, 21, 23, 28, 29, 32, 34, 36-38, 42	The most common batfish encountered, but only occasional sightings.	1-35
P. teira (Forsskål, 1775)	7, 13, 26	Occasional.	0-2
SCATOPHAGIDAE			
Scatophagus argus (Bloch, 1788)	1998-99		0-10
SIGANIDAE			
Siganus argenteus (Quoy and Gaimard, 1824)	2b, 2, 5, 6, 10, 11, 13-16, 18, 20, 22, 25, 26, 33, 37, 44	Moderately common.	1-30
S. canaliculatus (Park, 1797)	6, 11, 13, 41	Occasional in seagrass beds.	0-5

SPECIES	SITE RECORDS	ABUNDANCE	DEPTH (m)
S. corallinus (Valenciennes, 1835)	1, 2a, 6, 11-13, 15-18, 20-22, 24-26, 28, 30, 33, 34, 36, 37, 39, 41-44	Moderately common.	4-25
S. guttatus (Bloch, 1787)	1998-99		1-25
S. javus (Linnaeus, 1766)*	2a, 4, 5, 7, 14, 21	Occasional.	1-25
S. lineatus (Linnaeus, 1835)	4-6, 8, 16, 21, 38, 39, 44	Occasional.	1-25
S. puellus (Schlegel, 1852)	1-3, 6, 7, 10, 11, 13-18, 20-22, 24-26, 29, 30-32, 34, 37, 42-44	Common.	2-30
S. punctatissimus Fowler and Bean, 1929	1-3, 5, 6, 10, 11, 13, 15, 18, 20-22, 25-27, 29, 32, 35-38, 42	Moderately common.	3-30
S. punctatus (Forster, 1801)	1998-99		1-40
S. spinus (Linnaeus, 1758)	11	Rare.	1-12
S. virgatus (Valenciennes, 1835)	2a, 2b, 3-6, 11, 13, 15-17, 19, 20, 24, 25, 29, 33, 35, 38, 40-42	Moderately common.	1-20
S. vulpinus (Schlegel and Müller, 1844)	1, 2a, 2b, 5, 6, 10-17, 20-29, 31, 35, 37, 38, 40-44	Moderately common.	1-30
ZANCLIDAE			
Zanclus cornutus Linnaeus, 1758*	1-3, 5-6, 9-26, 28-34, 36-44	Common.	1-180
ACANTHURIDAE			
Acanthurus bariene Lesson, 1830*	9, 14, 33, 36	Occasional. Waigeo is type locality.	15-50
A. blochi Valenciennes, 1835	1, 2b, 14, 19, 21, 24, 26, 29	Occasional.	3-20
A. fowleri de Beaufort, 1951	1, 3, 16, 18, 25, 26	Occasional.	10-30
A. leucocheilus Herre, 1927	7, 15, 17, 18, 21, 22, 28, 31, 39	Occasional.	5-20
A. lineatus (Linnaeus, 1758)	1, 10, 11, 13, 15-17, 20-22, 24-26, 28, 30-34, 36, 40	Moderately common, usually in shallow surge-affected areas.	1-15
A. maculiceps (Ahl, 1923)	11, 14, 16, 20, 30-33, 36, 37, 42	Occasional.	1-15
A. mata (Cuvier, 1829)	2a, 6, 7, 9, 11, 14-18, 20-22, 25, 26, 28, 29, 30-34, 36, 39	Moderately common, usually on dropoffs in turbid water.	5-30

SPECIES	SITE RECORDS	ABUNDANCE	DEPTH (m)
A. nigricans (Linnaeus, 1758)*	2a, 7, 14, 18, 20, 26, 30-34, 36, 43	Occasional.	3-65
A. nigricaudus Duncker and Mohr, 1929	1, 2a, 3, 5-7, 10, 11, 14-18, 20-26, 30-39, 44	Moderately common.	3-30
A. nigrofuscus (Forsskål, 1775)	30-32	Occasional, but easily overlooked.	2-20
A. olivaceus Bloch and Schneider, 1801	1, 7, 11, 13-15, 17, 20, 25, 28, 30-34, 36, 43	Occasional.	5-45
A. pyroferus Kittlitz, 1834	1-3, 6, 7, 10, 11, 13, 14, 16-18, 20-22, 24-26, 28, 30-34, 36-39, 41, 44	Common.	4-60
A. thompsoni (Fowler, 1923)	1, 2a, 2b, 7, 10, 13-17, 20, 22, 36, 37, 39, 44	Moderately common, usually on steep dropoffs.	4-75
A. triostegus (Linnaeus, 1758)	1, 2a, 11	Occasional, usually in shallow wave-affected areas.	0-90
A. xanthopterus Valenciennes, 1835	1-5, 8, 9, 13, 15, 21, 23, 31, 40, 41, 44	Occasional, usually on sandy slopes adjacent to reefs.	3-90
Ctenochaetus binotatus Randall, 1955	1, 2b, 3, 6, 7, 9-18, 20, 22, 25, 26, 28-35, 37-44	Common.	10-55
C. striatus (Quoy and Gaimard, 1824)	1-3, 7, 11-16, 19-22, 24-26, 28, 30-37, 39-44	Common, usually in depths less than 10 m.	2-30
C. strigosus (Bennett, 1828)	2a, 30, 31	Only a few noticed, but hard to differentiate from C. striatus at a distance.	3-25
C. tominiensis Randall, 1955	26, 40, 43, 44	Occasional.	5-40
Naso annulatus (Quoy and Gaimard, 1825)	28	Rare, less than 10 seen.	15-40
N. brachycentron (Valenciennes, 1835)*	25, 28, 36	Rare, about 8 seen. Waigeo is type locality.	15-50
N. brevirostris (Valenciennes, 1835)	7, 14, 21, 22, 37, 39	Occasional, but common at site 14.	4-50
N. caeruleacauda Randall, 1994	2a, 39	Two schools seen.	5-40
N. hexacanthus (Bleeker, 1855)	1, 7, 10, 11, 14-18, 20-22, 26, 28, 31-34, 36, 37, 39, 42, 43	Moderately common.	6-140

SPECIES	SITE RECORDS	ABUNDANCE	DEPTH (m)
N. lituratus (Bloch and Schneider, 1801)	1, 2a, 6, 7, 10, 11, 13, 14, 16-22, 24-26, 29-36, 39, 41-44	Common.	5-90
N. lopezi Herre, 1927	7, 13, 14, 16-18, 26, 28, 30-34, 39, 42	Moderately common.	6-70
N. minor (Smith, 1966)	25	One school of about 30 seen.	10-50
N. thymnoides (Valenciennes, 1835)	7, 20, 28, 30	Occasional large schools seen.	8-50
N. unicornis Forsskål, 1775*	16, 25, 33, 37, 41	Occasional.	4-80
N. vlamingii Valenciennes, 1835	1, 6, 7, 10, 14, 16-18, 20-22, 25, 33, 36, 37, 40, 42	Moderately common, adjacent to steeper outer slopes.	4-50
Paracanthurus hepatus (Linnaeus, 1758)*	2a, 7, 14, 22, 25, 32, 34	Occasional, but common at sites 14 and 32.	2-40
Zebrasoma scopas Cuvier, 1829	1-3, 6, 7, 10-22, 24-26, 28, 30-44	Common.	1-60
Z. veliferum Bloch, 1797*	2a, 2b, 5-22, 24-27, 29-32, 36-42, 44	Common.	4-30
SPHYRAENIDAE			
Sphyraena barracuda (Walbaum, 1792)*	13, 41	Rare, only 2 seen.	0-20
S. flavicauda Rüppell, 1838*	18, 26	Rare, 2 schools of about 10-30 fish seen.	1-20
S. jello Cuvier, 1829	13, 14	Rare, only 2 seen.	1-20
S. qenie Klunzinger, 1870	1998-99	1	5-40
SCOMBRIDAE			
Euthynnus affinis (Cantor, 1849)	1998-99		0-20
Grammatorcynus bilineatus (Quoy and Gaimard, 1824)	7	Rare, only one seen.	10-40
Gymnosarda unicolor (Rüppell, 1836)	1998-99		5-100
Rastrelliger kanagurta (Cuvier, 1816)*	1998-99		0-30
Scomberomorus commerson (Lacepède, 1800)	16, 18, 43	Rare, 3 large individuals seen.	0-30
BOTHIDAE			
Bothus mancus Broussonet, 1782*	1998-99		5-30

SPECIES	SITE RECORDS	ABUNDANCE	DEPTH (m)
B. pantherinus (Rüppell, 1830)*	1998-99		5-30
SOLEIDAE			
Soleichthys heterorhinos (Bleeker, 1856)*	1998-99		2-25
BALISTIDAE			
Balistapus undulatus (Park, 1797)*	1-3, 5-18, 20-22, 24-44	Common.	3-50
Balistoides conspicillum (Bloch and Schneider, 1801)	1, 2a, 3, 7, 10, 13, 14, 16, 18, 20, 21, 24-26, 28, 30-33, 36, 37, 39	Common.	10-50
B. viridescens (Bloch and Schneider, 1801)	2b, 3, 5-8, 13-16, 18, 21-25, 28, 29, 37, 39, 44	Occasional.	5-45
Canthidermis maculatus (Bloch, 1786)	34	Rare, collected from floating *Sargassum*.	1-30
Melichthys vidua (Solander, 1844)	1, 2a, 14, 16, 17, 20, 22, 25, 28, 30, 32, 36, 37	Occasional.	3-60
Odonus niger Rüppell, 1836*	1, 2a, 3, 6, 7, 10-17, 20, 25, 28, 30-34, 36, 37, 39, 42-44	Common.	3-40
Pseudobalistes flavimarginatus (Rüppell, 1828)*	3-6, 9, 10, 13, 14, 23, 25, 29, 34, 38	Occasional, in sheltered sand or rubble areas.	2-50
P. fuscus (Bloch & Schneider, 1801)	37	Rare, only one seen.	1-50
Rhinecanthus aculeatus (Linnaeus, 1758)*	12	Rare, only one seen.	0-3
R. rectangulus (Bloch and Schneider, 1801)*	31, 33, 36	Rare, less than 10 seen.	1-3
R. verrucosus (Linnaeus, 1758)*	2a, 6, 11-13, 15, 16, 24, 41	Occasional, but locally common on shallow flats near shore.	0-3
Sufflamen bursa (Bloch and Schneider, 1801)	1, 2a, 3, 6, 7, 10, 11, 13-15, 17, 18, 20, 22, 24-26, 28, 29-37, 39, 41-44	Common.	3-90
S. chrysoptera (Bloch and Schneider, 1801)*	1-3, 6, 7, 9-17, 20-22, 24-26, 28-34, 36-39, 42	Moderately common.	1-35
S. fraenatus (Latreille, 1804)	2a, 13, 15, 28, 30, 33, 42	Occasional.	8-185

SPECIES	SITE RECORDS	ABUNDANCE	DEPTH (m)
MONACANTHIDAE			
Acreichthys tomentosus (Linnaeus, 1758)*	1998-99		1-10
Aluterus scriptus (Osbeck, 1765)*	2b, 16, 18, 25	Circumtropical. Rare, only 4 observed.	2-80
Amanses scopas Cuvier, 1829	7, 10, 11, 13, 21, 25, 27	Occasional.	3-20
Cantherines dumerilii Hollard, 1854	26	Rare, only one seen.	1-35
C. fronticinctus (Günther, 1866)	1, 11, 13, 15, 16, 18, 20, 25, 28, 32, 34, 36, 42, 43	Occasional.	2-40
C. pardalis (Rüppell, 1866)	17, 26, 28	Rare, only 5 seen.	2-20
Oxymonacanthus longirostris Bloch and Schneider, 1801	2b	Rare, a single pair seen.	1-30
Paraluteres prionurus (Bleeker, 1851)	12	Rare, only one seen.	2-25
Paramonacanthus japonicus (Tilesius, 1801)	1998-99		2-30
Pervagor janthinosoma (Bleeker, 1854)	20, 36	Rare, only 2 seen.	2-18
P. melanocephalus (Bleeker, 1853)	13	Rare, only one seen.	15-40
P. nigrolineatus (Herre, 1927)	1998-99		2-15
Pseudomonacanthus macrurus (Bleeker, 1856)	13	Rare, only one seen. Photographed.	5-40
OSTRACIIDAE			
Ostracion cubicus Linnaeus, 1758*	2b, 12, 14, 15, 18, 20, 25, 28, 34, 38	Occasional.	1-40
O. meleagris Shaw, 1796*	7, 11, 16, 25, 30, 33, 34, 37, 43	Occasional.	2-30
O. solorensis Bleeker, 1853*	6, 11, 16	Rare, only 3 seen.	1-20
TETRAODONTIDAE			
Arothron caeruleopunctatus Matsuura, 1994	18	Rare, only one seen.	5-30
A. hispidus (Linnaeus. 1758)*	20, 24, 38	Rare, only 3 seen.	1-50
A. manilensis (Marion de Procé, 1822)*	1998-99		1-20
A. mappa (Lesson, 1830)	4, 5, 14, 18, 21, 34, 36, 39	Occasional.	4-40
A. nigropunctatus (Bloch and Schneider, 1801)*	1, 6, 7, 10, 11, 13, 14, 16, 18, 21, 23, 25, 26, 29, 32, 34, 37-39, 41, 43, 44	Moderately common, but always in low numbers at each site.	2-35

SPECIES	SITE RECORDS	ABUNDANCE	DEPTH (m)
A. stellatus (Schneider, 1801)	1998-99		3-58
Canthigaster amboinensis (Bleeker, 1865)	1	Rare, only one seen.	0-5
Canthigaster bennetti (Bleeker, 1854)	1998-99		1-10
C. compressa (Procé, 1822)	1998-99		1-20
C. janthinoptera (Bleeker, 1855)	1998-99		9-60
C. papua Bleeker, 1848*	2b, 3, 11, 15, 29, 35, 43	Occasional.	1-20
C. valentini (Bleeker, 1853)	2a, 3, 11, 12, 14-16, 18, 20, 25, 28-31, 33, 41, 42	Occasional.	3-55
DIODONTIDAE			
Diodon hystrix Linnaeus, 1758	2a, 14, 34	Rare, only 3 seen.	1-30
D. liturosus Shaw, 1804	1998-99		3-35

Appendix 5

List of target (commercially important) fishes of the Raja Ampat Islands

La Tanda

SPECIES	SITE RECORDS
HOLOCENTRIDAE	
Myripristis adusta Bleeker, 1853	3, 7, 14, 16, 17, 33, 34, 39, 41
Myripristis berndti (Jordan & Evermann, 1903)	7, 15, 33, 34, 39, 41, 42, 44
Myripristis botche Cuvier, 1829	14
Myripristis kuntee Valenciennes, 1831	4, 6, 8, 11, 14, 16, 21, 22, 24, 30, 34
Myripristis murdjan (Forsskål, 1775)	3, 1-17, 19, 21, 25, 28, 31
Myripristis violacea Bleeker, 1851	2a, 2b, 11, 14, 16, 25, 26, 28, 31, 41, 42, 44
Myripristis sp.	2b, 14, 19, 33, 36, 43
Sargocentron spiniferum (Forsskål, 1775)	1, 2a, 2b, 6,-9, 12, 14, 16, 21-23, 28, 29, 38, 39, 41
Sargocentron caudimaculatum (Rüppell, 1838)	1, 7, 14, 17, 21, 24, 25, 30, 33, 34, 38, 39, 42, 43
Sargocentron spiniferum (Forsskål, 1775)	1, 2a, 2b, 6,-9, 12, 14, 16, 21-23, 28, 29, 38, 39, 41
Neoniphon argenteus (Valenciennes, 1831)	14
Neoniphon sammara (Forsskål, 1775)	2b, 3, 35, 44
SERRANIDAE	
Aetalopherca rogaa (Forsskål, 1775)	2a, 12, 15, 17, 21, 22, 25, 30, 33, 43, 44
Anyperodon leucogrammicus (Valenciennes, 1828)	15, 17, 18, 20, 29, 35, 43
Cephalopolis argus Bloch & Schneider, 1801	1, 2a, 7, 17, 20, 33, 34, 36-38, 42, 44
Cephalopolis boenack Bloch, 1790	4, 6, 7, 10, 20, 24, 27, 28, 29, 33, 34, 43
Cephalopolis cyanostigma (Valenciennes, 1828)	1, 2a, 3, 4, 6, 7, 11, 14, 16, 19, 20, 23, 24, 25, 26, 28, 29, 33, 37, 40-44
Cephalopolis leopardus (Lacepéde, 1801)	1, 2a, 3, 7, 22, 25, 33
Cephalopolis microprion (Bleeker, 1852)	2b, 6, 8, 12, 16, 19, 26, 27, 29, 38, 40-44
Cephalopolis miniata (Forsskål, 1775)	1, 2a, 3, 10, 12, 14-18, 20-22, 25, 28-30, 42, 43
Cephalopolis sexmaculata (Rüppell, 1830)	25
Cephalopolis sonnerati (Valenciennes, 1828)	4, 10, 12, 20-25, 28, 38, 39
Cephalopolis urodeta (Schneider, 1801)	1, 2a, 3, 7, 10, 13,14, 20, 24, 25, 28, 30-33, 34, 36, 37
Cephalopolis sp.	41
Epinephelus fasciatus (Forsskål, 1775)	14, 15, 17, 18, 20
Epinephelus maculatus (Bloch, 1790)	15, 22, 19
Epinephelus merra Bloch, 1793	37, 39
Epinephelus hexagonatus (Bloch & Schneider, 1801)	1, 2a, 27
Epinephelus ongus (Bloch, 1790)	28, 32, 35
Epinephelus sp.	4, 39
Gracilla albomarginata (Fowler & Bean, 1930)	1, 32
Plectropomus leopardus (Lacepéde, 1802)	2b, 13, 15, 18
Plectropomus maculatus (Bloch, 1790)	4, 8, 11-13, 16, 26, 29, 31, 35, 44
Plectropomus oligocanthus (Bleeker, 1854)	6, 16, 17, 27, 39, 40-44
Variola albomarginatus Baissac, 1953	1, 41
Variola louti (Forsskål, 1775)	2a, 10, 13, 21, 31, 33, 34, 36
Chromileptes altivelis (Valenciennes, 1828)	11, 16, 18, 20

SPECIES	SITE RECORDS
PRIACANTHIDAE	
Priacanthus hamrur (Forsskål, 1775)	30
Priacanthus sp.	13
CARANGIDAE	
Elegatis bipinnulata (Quoy & Gaimard, 1825)	6, 7
Caranx ignobilis (Forsskål, 1775)	42
Caranx lugubris Poey, 1861	33
Caranx melampygus Cuvier, 1833	2a, 4, 6, 14, 21
Caranx sexfasciatus Quoy & Gaimard, 1825	42
Caranx sp.	41-43
Carangoides bajad (Forsskål, 1775)	2a, 7, 12, 16, 18-20, 26, 28, 39, 41-43
Carangoides ferdau (Forsskål, 1775)	4, 6
Scomberoides lysan (Forsskål, 1775)	2a
Selaroides leptolepis (Kuhl & van Hasselt, 1833)	4, 23
LUTJANIDAE	
Aprion virescens Valenciennes, 1830	39
Lutjanus biguttatus (Valenciennes, 1930)	4, 6, 19, 23, 26, 27, 29, 31, 35, 38, 40-44
Lutjanus bohar (Forsskål, 1775)	1, 10, 13, 15, 17, 18, 22, 28, 38
Lutjanus carponotatus (Richardson, 1842)	3, 4, 6, 8, 10, 12, 15, 16, 27, 35, 39, 40, 42
Lutjanus decussatus (Cuvier, 1828)	3, 6, 7, 10, 11, 14, 15, 19, 20, 26, 27, 34, 35, 37, 39-44
Lutjanus ehrenbergi Peters, 1869	3
Lutjanus fulvus (Schneider, 1801)	1-4,7, 8, 11, 14, 19, 30, 31, 38, 44
Lutjanus gibbus (Forsskål, 1775)	8, 10-12, 16, 28, 29, 36, 38, 39, 43
Lutjanus kasmira (Forsskål, 1775)	28, 39
Lutjanus monostigma (Cuvier, 1828)	1, 2a, 3, 13, 14, 19, 39, 43, 44
Lutjanus quinquelineatus (Bloch, 1790)	8, 14, 25, 41, 44
Lutjanus rivulatus (Cuvier, 1828)	4, 8, 12, 39, 40
Lutjanus russelli (Bleeker, 1849)	4, 10, 19, 36
Lutjanus semicinctus Quoy & Gaimard, 1824	1, 2b, 3, 6, 8, 10, 13, 14, 16, 18-20, 22, 24-26, 31, 33, 36, 37, 39, 41
Lutjanus sp.	12, 15
Macolor macularis Fowler, 1931	1, 2a, 7, 10, 14,16-18, 39
Macolor niger Forsskål, 1775	10, 11, 13, 14, 16, 18, 28, 30-33, 36-38, 43, 44
Paracaesio sordidus Abe & Shinohora, 1962	36
Symphorichthys spilurus (Günther, 1874)	3
Symphorus nematophorus (Bleeker, 1860)	12, 18
CAESIONIDAE	
Caesio caerulaurea Lacepéde, 1801	1, 2a, 2b, 7, 11-15, 18, 19, 21, 23, 28, 33, 34, 36, 37, 39-41, 43
Caesio cuning (Bloch, 1791)	1-4, 6-9, 11-19, 21, 23, 26-29, 33, 35-38, 40-44
Caesio lunaris Cuvier, 1830	1, 2a, 2b, 7, 9, 11, 12, 14, 15, 25, 28, 31, 32, 36, 37, 39, 40, 44
Caesio teres, Seale, 1906	2b, 4, 14-16, 19, 20, 21, 25, 28, 32, 37, 39, 44
Caesio xanthonota Bleeker, 1853	3, 10, 20
Pterocaesio chrysozona (Cuvier, 1830)	16

SPECIES	SITE RECORDS
Pterocaesio diagramma (Bleeker, 1865)	21, 22
Pterocaesio marri Schultz, 1953	3, 13, 14, 16-18, 22, 25, 32-34, 36, 44
Pterocaesio pisang (Bleeker, 1853)	1, 2a, 8, 10-18, 20-22,25,29, 31-34, 40-44
Pterocaesio randalli Carpenter, 1987	1, 2a, 14, 39, 43
Pterocaesio tessellata Carpenter, 1987	14, 17, 18, 21, 32
Pterocaesio tile (Cuvier, 1830)	1, 7, 14, 25, 28, 32, 36, 37
Pterocaesio trilineata Carpenter, 1987	7
HAEMULLIDAE	
Plectorhinchus chaetodontoides Lacepéde, 1800	2a, 12, 16, 17, 19, 22, 26, 29, 36, 40, 41, 44
Plectorhinchus chrysotaenia (Bleeker, 1855)	30
Plectorhinchus gibbosus (Lacepéde, 1802)	9, 34, 38
Plectorhinchus lessoni (Cuvier, 1830)	17, 28, 30, 32, 34-36, 39, 42
Plectorhinchus lineatus (Linnaeus, 1758)	1, 2a, 3, 7, 10, 11, 15, 16, 25, 30, 40, 44
Plectorhinchus picus (Cuvier,1830)	12, 40, 44
Plectorhinchus polytaenia (Bleeker, 1852)	1, 2a, 3, 7-10, 12-16, 18, 21, 22, 25, 26, 28, 30, 37-39, 42-44
Plectorhinchus vittatus (Linnaeus, 1758)	1, 3, 10, 14, 28, 33, 39
LETHRINIDAE	
Lethrinus erythropterus Valenciennes, 1830	1, 2a, 3, 6, 7, 11-16, 19, 20, 25, 26, 29, 33, 35, 39, 43, 44
Lethrinus harak (Forsskål, 1775)	11, 33
Lethrinus nebulosus (Forsskål, 1775)	3
Lethrinus obsoletus (Forsskål, 1775)	7, 11, 18, 24
Lethrinus olivaceus Valenciennes, 1830	24, 28, 44
Lethrinus sp.	
Monotaxis grandoculus (Forsskål,1775)	1-3, 7, 10, 11, 13, 14, 17, 18, 20, 21, 25, 26, 28, 29, 31-34, 36-40, 42, 44
NEMIPTERIDAE	
Scolopsis affinis Peters, 1877	24, 30
Scolopsis bilineatus (Bloch, 1793)	1, 3, 7, 9, 10, 12, 14-18, 20-22, 24, 25, 28, 30-34, 36, 37, 39, 40, 42-44
Scolopsis ciliatus (Lacepéde, 1802)	3, 4, 9, 23, 40
Scolopsis lineatus Quoy & Gaimard, 1824	2b
Scolopsis margaritifer (Cuvier, 1830)	1, 2b, 3, 4, 6, 8, 11-13, 15, 16, 23, 24, 26, 27, 29-31, 35, 37, 38, 40
Scolopsis monogramma (Kuhl & van Hasselt, 1830)	4, 12, 21, 41-44
Scolopsis trilineatus Kner, 1868	2b, 22, 44
Scolopsis xenochrous Günther, 1872	24
Scolopsis sp.	37.
Pentapodus caninus (Cuvier,1830)	3, 7, 12, 15, 21, 24, 27, 31
Pentapodus emeryi (Richardson, 1843)	3, 7, 12, 14, 15, 21, 29, 31, 33, 38
Pentapodus trivittatus (Bloch, 1791)	3, 4, 12, 15, 19, 23, 26, 29, 40
MULLIDAE	
Parupeneus barberinus (Lacepéde, 1801)	2b, 3, 4, 7-12, 14-31, 37-44
Parupeneus bifasciatus (Lacepéde, 1801)	1, 2a, 3, 6, 7, 10-12, 14-17, 20, 21, 25, 28, 31, 33, 34, 37, 40, 42, 43
Parupeneus cyclostomus (Lacepéde, 1801)	1, 2a, 3, 7, 10, 14, 15, 18, 20, 24, 25, 31, 39-44

SPECIES	SITE RECORDS
Parupeneus macronema (Lacepéde, 1801)	30.
Parupeneus multifasciatus (Quoy & Gaimard, 1825)	1-3, 6, 7, 10-16, 18, 20-44
Parupeneus pleurostigma (Bennett,1830)	30, 37, 40
Mulloides flavolineatus (Lacepéde, 1801)	12, 13, 29
Mulloides vanicolensis (Valenciennes, 1831)	26, 31
Upeneus tragula Richardson, 1846	4, 15, 16, 18, 29, 40

KYPHOSIDAE

Kyphosus vaigiensis (Quoy & Gaimard, 1825)	1, 2a, 16, 17, 20, 24, 39

MONODACTYLIDAE

Monodactylus argenteus (Linnaeus, 1758)	8, 26

LABRIDAE

Cheilinus fasciatus (Bloch, 1791)	1-4, 6, 7, 9, 11-19, 21, 23-27, 29-31, 33, 35, 37-44
Cheilinus trilobatus Lacepéde, 1802	15, 16, 18, 25, 28, 33, 34, 40
Cheilinus undulatus Rüppell, 1835	39, 40
Cheilio inermis (Forsskål, 1775)	15
Choerodon anchorago (Bloch, 1791)	2a, 2b, 3, 11-13, 15, 19, 35, 38, 40, 43, 44
Epibulus insidiator (Pallas, 1770)	17, 27, 37
Hemigymnus fasciatus(Bloch, 1792)	2b, 21, 26, 34, 37
Hemigymnus melapterus (Bloch, 1791)	13, 17, 24, 35, 37, 39, 42, 44
Oxycheilinus bimaculatus Valenciennes, 1840	2b
Oxycheilinus celebicus Bleeker, 1853	2b

SCARIDAE

Bolbometopon muricatum (Valenciennes, 1840)	2a, 13, 15, 21, 28, 33, 38
Cetoscarus bicolor (Rüppell, 1828)	2b, 6, 7, 10, 12-14, 17, 18, 25, 26, 29, 31, 32, 35, 38-42, 44
Chlorurus bleekeri (de Beaufort, 1940)	1-3, 6, 7, 10, 11, 14-22, 24-27, 29-32, 34, 35, 37-44
Chlorurus bowersi (Snyder, 1909)	2b, 11-14, 16, 21, 24, 26, 27, 42
Chlorurus sordidus (Forsskål, 1775)	1-3, 6, 8, 10-15, 17, 19, 20, 24-27, 29, 31, 35, 37, 38-42
Chlorurus microrhinos (Bleeker, 1854)	43, 44
Hipposcarus longiceps (Bleeker, 1862)	2b, 10, 12-14, 38, 41, 42, 44
Scarus flavipectoralis Schultz, 1958	1, 2b, 11-15, 19, 20, 24, 27, 35, 37-44
Scarus forsteni (Bleeker, 1861)	17, 20, 22, 24-26, 32-34, 36, 39, 44
Scarus frenatus Lacepéde, 1802	20, 22, 29, 32, 34, 37, 39, 41, 42
Scarus ghobban Forsskål, 1775	2a, 6, 10, 15, 18, 23, 25, 30-32, 44
Scarus hypselopterus Bleeker, 1853	26, 37, 41
Scarus niger Forsskål, 1775	16, 17, 21, 25, 26, 35, 37-44
Scarus oviceps Valenciennes, 1839	7, 25
Scarus quoyi Valenciennes, 1840	17, 42
Scarus rivulatus Valenciennes, 1840	16, 18-20, 24, 26, 27, 29, 38, 40-44
Scarus rubroviolaceus Bleeker, 1849	1, 12, 14, 16, 17, 20-22, 25, 26, 29, 30-34, 36-44
Scarus schlegeli (Bleeker, 1861)	1, 2a, 3, 10, 14, 16, 18, 22, 32, 44
Scarus tricolor Bleeker, 1847	32, 39
Scarus sp	2a, 10, 13-15, 17-20, 22-24, 26, 27, 31, 39, 42, 44

EPHIPPIDAE

Platax orbicularis (Forsskål, 1775)	10, 13, 14, 17, 39

SPECIES	SITE RECORDS
Platax pinnatus (Linnaeus, 1758)	29, 42
Platax teira (Forsskål, 1775)	39, 42
SIGANIDAE	
Siganus argenteus (Quoy & Gaimard, 1825)	2b, 10, 14, 16, 17, 24, 31, 32, 34, 41
Siganus canaliculatus (Park, 1797)	1, 2a, 6, 11, 36
Siganus corallinus (Valenciennes, 1835)	2a, 2b, 6, 11, 13-20, 22, 25, 26, 28, 31, 33, 34, 37-40, 42-44
Siganus doliatus (Cuvier, 1830)	1, 2b, 11, 14-17, 19, 21, 26, 27, 29, 34, 35, 38, 40
Siganus fuscescens (Houttyn, 1782)	6, 11
Siganus guttatus (Bloch, 1787)	16
Siganus javus (Linnaeus, 1766)	4
Siganus lineatus (Linnaeus, 1835)	4, 7-9, 13, 16, 21, 29, 38, 39, 41, 42
Siganus puellus Schlegel, 1852	2a, 2b, 3, 7-11, 14, 16-18, 21, 22, 25, 26, 29, 31, 37, 38, 43, 44
Siganus punctatissimus (Forster, 1801)	1a, 2a, 11-16, 18-20, 25-27, 35, 37, 38, 41-44
Siganus vulpinus (Schlegel & Muller, 1845)	1, 2b, 6, 10-12, 16, 20, 25-27, 31, 32, 35, 37, 39-42, 44
ACANTHURIDAE	
Acanthurus bariene Lesson, 1830	7, 27
Acanthurus blochii Valenciennes, *1835*	2-4, 6, 8-15, 20, 27, 28, 30-36, 40, 42-44
Acanthurus leucocheilus Herre, 1927	16-18, 20-22, 26
Acanthurus lineatus (Linnaeus, 1758)	2a, 3, 6, 15, 16, 20, 25, 28, 30-34, 39, 41
Acanthurus mata (Cuvier, 1829)	7, 10, 11, 14-18, 20-22, 25, 28, 32, 33, 36, 39, 42
Acanthurus nigricans (Linnaeus, 1758)	2a, 7, 13, 14, 17, 18, 28, 32, 34, 36, 37
Acanthurus nigricauda Duncker & Mohr, 1929	14, 32
Acanthurus nubilus (Fowler and Bean, 1929)	39
Acanthurus olivaceus Bloch & Schneider, 1801	11, 14, 20, 21, 30, 32, 33, 34
Acanthurus pyroferus Kittlitz, 1834	1-3, 6, 7, 10, 11, 13, 14, 16-18, 20, 21, 24-26, 30-33, 36-39,41, 43, 44
Acanthurus thompsoni (Fowler, 1923)	1, 2a, 2b, 7, 10, 13, 14, 17, 22, 32, 36, 39, 42, 44
Acanthurus xanthopterus Valenciennes, 1835	10, 15, 25, 32, 39
Acanthurus sp.	44
Ctenochaetus binotatus Randall, 1955	1, 2a, 3, 7, 10, 11, 13-16, 21, 24, 25, 30, 31, 32, 34, 35, 38-41, 44
Ctenochaetus striatus (Quoy & Gaimard, 1825)	1-3, 6, 7, 10-12, 14-22, 24-26, 29-44
Ctenochaetus tominiensis Randall, 1955	16, 22, 26, 31, 40, 44
Paracanthurus hepatus (Linnaeus, 1766)	14, 25, 32
Naso brachycentron (Valenciennes, 1835)	36
Naso brevirostris (Valenciennes, 1835)	14, 16-18, 33, 37
Naso hexacanthus (Bleeker, 1855)	1, 2a, 6, 7, 10, 12-19, 25, 28, 30, 33, 34, 36, 37, 39, 40, 42-44
Naso lituratus (Bloch & Schneider, 1801)	1, 2a, 2b, 6, 10, 13, 14, 17, 21, 22, 25, 26, 28, 30, 31, 33, 36
Naso lopezi Herre, 1927	18, 28, 32, 33, 39
Naso thynnoides (Valenciennes, 1835)	7, 31, 42, 43
Naso unicornis Forsskål, 1775	14, 25, 27

SPECIES	SITE RECORDS
Naso vlamingii (Valenciennes, 1835)	2a, 14, 16, 18, 20, 26, 28, 42
Zebrasoma scopas Cuvier, 1829	1, 2b, 3, 6, 7, 10, 12-17, 20-22, 24-26, 30-32, 35, 37-39, 41-44
Zebrasoma veliferum Bloch, 1797	1, 2a, 2b, 6, 8, 11, 14, 15, 19, 21, 22, 26, 27, 31, 37, 39
SPHYRAENIDAE	
Sphyraena jello Cuvier, 1829	26
Sphyraena qenie Klunzinger, 1870	15
SCOMBRIDAE	
Rastrelliger kanagurta (Cuvier, 1816)	13, 44
Scomberomorus commerson (Lacepéde, 1800)	15, 39, 43

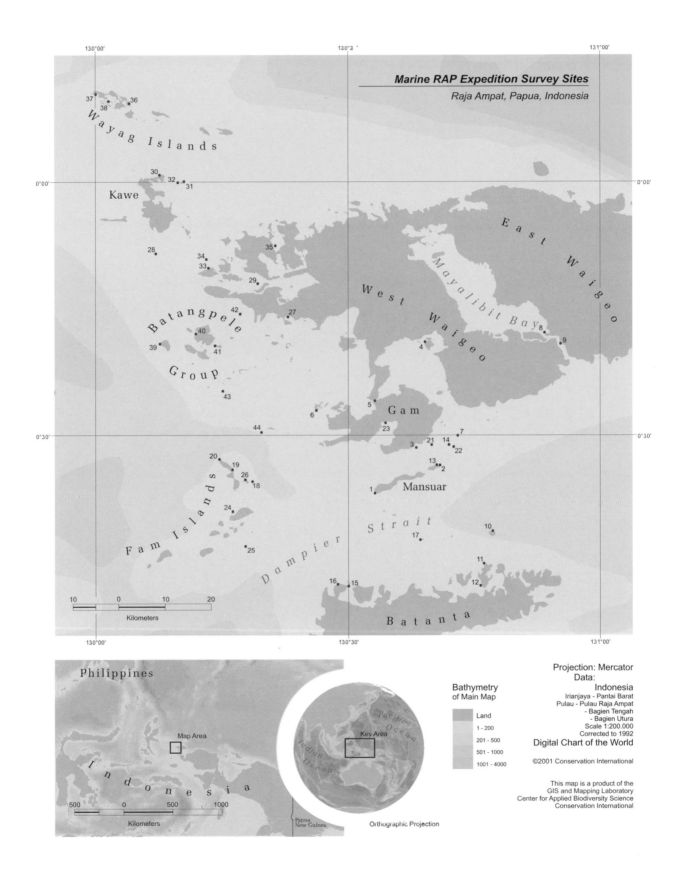

Marine RAP Expedition Survey Sites
Raja Ampat, Papua, Indonesia

130°00' 130°3' 131°00'

Wayag Islands

37 38 36

Kawe

30 32 31

East Waigeo

0°00' 0°00'

28

35

34
33

29

Mayalibit Bay

West Waigeo

8
9

Batangpele

42 27

40 4

39 41

Group

43

5

6 Gam

44 23

0°30' 3 21 14 7 0°30'
22

20 13 2

19 26
18 1 Mansuar

24

Fam Islands

25 10

Strait 17

11

16 15 12

Dampier

Batanta

10 0 10 20
Kilometers

130°00' 130°30' 131°00'

Philippines

Map Area

Pacific
Ocean

Key Area

Indian
Ocean

Indonesia

Bathymetry
of Main Map

Land
1 - 200
201 - 500
501 - 1000
1001 - 4000

500 0 500 1000
Kilometers

Papua
New Guinea

Orthographic Projection

Projection: Mercator
Data:
Indonesia
Irianjaya - Pantai Barat
Pulau - Pulau Raja Ampat
- Bagien Tengah
- Bagien Utura
Scale 1:200.000
Corrected to 1992
Digital Chart of the World

©2001 Conservation International

This map is a product of the
GIS and Mapping Laboratory
Center for Applied Biodiversity Science
Conservation International